NATURAL DISASTERS AND ADAPTATION TO CLIMATE CHANGE

Every year extreme events occur that are costly in terms of human life and damage to infrastructure. The extent of this damage is a measure of the extent to which adaptation, whether planned or autonomous, has been successfully undertaken. Extreme events, their impacts, and the measures undertaken to protect ecosystems, human activities, and human welfare contain instructive lessons for adaptation to the effects of climate change.

This volume presents eighteen case studies of natural disasters from Australia, Europe, North America, and developing countries. By comparing the impacts, it seeks to identify what moves people to adapt, which adaptive activities succeed and which fail, the underlying reasons for success or failure, and the factors that determine when adaptation is required and when simply bearing the impact may be the more appropriate response. Much has been written about the theory of adaptation and high-level, especially international, policy responses to climate change. This book aims to inform actual adaptation practice – what works, what does not, and why. It explores some of the lessons we can learn from past disasters and the adaptation that takes place after the event in preparation for the next. How successful are these actions, how do communities respond and behave, what makes an action fail, and how do these actions fare in the long term? As a modern society, we have no past experience of adapting to climate change, so managing and responding to historical extremes provides the only empirical evidence, however imperfect, of how successfully we may adapt to climate change now and in the future.

This volume will be especially useful for researchers and decision makers in policy and government concerned with climate change adaptation, emergency management, disaster risk reduction, and environmental policy and planning.

SARAH BOULTER is a Research Fellow with the National Climate Change Adaptation Research Facility (NCCARF), Griffith University, Gold Coast, Australia, where she works on synthesis and communication of adaptation research. She has been involved in the development of policy guidance development programs for adaptation, research programs on historical case studies, assessment of forest vulnerability in Australasia, and as the convenor of Australia's Climate Adaptation conferences. She is a contributing author to the Australia chapter of the Intergovernmental Panel on Climate Change Fifth Assessment Report. Her research background includes studies of biodiversity and reproductive ecology of forested systems and the impacts of climate change.

JEAN PALUTIKOF is the Director of the NCCARF, Griffith University, Gold Coast, Australia. At NCCARF, she has built a national program of adaptation research, communication, and partnerships. Her work at NCCARF has convinced her of the need for case studies of good practice in adaptation action, to build adaptive capacity and knowledge among decision makers. Prior to joining NCCARF, Professor Palutikof managed the production of the Intergovernmental Panel on Climate Change Fourth Assessment Report for Working Group II (Impacts, Adaptation and Vulnerability), while based at the UK Met Office. Prior to joining the Met Office, she was a Professor in the School of Environmental Sciences and Director of the Climatic Research Unit at the University of East Anglia, United Kingdom. Her research interests focus on climate change impacts and adaptation and the application of climatic data to economic and planning issues.

DAVID JOHN KAROLY is a Professor of Climate Science in the School of Earth Sciences at the University of Melbourne. He is an internationally recognized expert in climate change and climate variability, including greenhouse climate change, stratospheric ozone depletion, and interannual climate variations stemming from the El Niño–Southern Oscillation. He was heavily involved in preparation of the Fourth Assessment Report of the Intergovernmental Panel on Climate Change. He joined the University of Melbourne in 2007 as an ARC Federation Fellow funded by the Australian government. He is a member of the new Climate Change Authority in Australia, the Science Advisory Panel to the Australian Climate Commission, and the Wentworth Group of Concerned Scientists.

DANIELA GUITART is an environmental scientist working at the NCCARF, Griffith University, Gold Coast, Australia. At NCCARF she coordinates the Adaptation Research Network activities and manages the production of information tools that communicate climate change adaptation research. Prior to joining NCCARF, she conducted research on community-based urban agriculture and its contribution to food security and the conservation of agricultural biodiversity.

NATURAL DISASTERS AND ADAPTATION TO CLIMATE CHANGE

Edited by

SARAH BOULTER
National Climate Change Adaptation Research Facility, Griffith University

JEAN PALUTIKOF
National Climate Change Adaptation Research Facility, Griffith University

DAVID JOHN KAROLY
University of Melbourne

DANIELA GUITART
National Climate Change Adaptation Research Facility, Griffith University

CAMBRIDGE
UNIVERSITY PRESS

CAMBRIDGE
UNIVERSITY PRESS

University Printing House, Cambridge CB2 8BS, United Kingdom

One Liberty Plaza, 20th Floor, New York, NY 10006, USA

477 Williamstown Road, Port Melbourne, VIC 3207, Australia

4843/24, 2nd Floor, Ansari Road, Daryaganj, Delhi - 110002, India

79 Anson Road, #06-04/06, Singapore 079906

Cambridge University Press is part of the University of Cambridge.

It furthers the University's mission by disseminating knowledge in the pursuit of education, learning and research at the highest international levels of excellence.

www.cambridge.org
Information on this title: www.cambridge.org/9781108445979

© Cambridge University Press 2013

First published 2013
First paperback edition 2017

A catalogue record for this publication is available from the British Library

Library of Congress Cataloging in Publication data
Natural disasters and adaptation to climate change / [compiled by] Sarah
Boulter, Griffith University, Jean Palutikof, Griffith University, David John
Karoly, University of Melbourne, Daniela Guitart, Griffith University.
pages cm
Includes bibliographical references and index.
ISBN 978-1-107-01016-1 (hardback)
1. Storms. 2. Emergency management. 3. Disaster relief. 4. Human beings – Effect of
climate on. I. Boulter, Sarah, 1970– editor of compilation. II. Palutikof, J. P. editor of
compilation. III. Karoly, David John, 1955– editor of compilation. IV. Guitart,
Daniela, 1988– editor of compilation.
HV635.5.N38 2014
363.34–dc23   2013019454

ISBN 978-1-107-01016-1 Hardback
ISBN 978-1-108-44597-9 Paperback

Contents

Part III Case Studies from Europe

Part IV Case Studies from the Developing World

Part V Synthesis Chapters

Contributors

W. Neil Adger, College of Life and Environmental Sciences, University of Exeter

Jeroen Aerts, Institute for Environmental Studies, VU University Amsterdam

Armando Apan, Australian Centre for Sustainable Catchments, University of Southern Queensland

Jessica Ayers, International Institute for Environment and Development

Jon Barnett, Department of Resource Management and Geography, University of Melbourne

Juan F. Barrera, El Colegio de la Frontera Sur, Tapachula, Chiapas

Simon P. J. Batterbury, Department of Resource Management and Geography, University of Melbourne

Linda C. Botterill, Faculty of Business, Government and Law, University of Canberra

Sarah Boulter, National Climate Change Adaptation Research Facility, Griffith University

Edwin Castellanos, Centro de Estudios Ambientales y de Biodiversidad, Universidad del Valle de Guatemala

Declan Conway, School of International Development, University of East Anglia

Gustavo Cruz-Bello, Centro de Investigación en Geografía y Geomática 'Ing. Jorge L. Tamayo'

W. Priyan S. Dias, Department of Civil Engineering, University of Moratuwa

Markus G. Donat, Climate Change Research Centre, University of New South Wales

Stephen Dovers, Australian National University

Thomas E. Downing, Global Climate Adaptation Partnership

Hallie Eakin, School of Sustainability, Arizona State University

C. J. Fotheringham, US Geological Survey, Western Ecological Research Center

Andrew W. Garcia (Deceased), Coastal and Hydraulics Laboratory Engineer Research and Development Center, US Army Corps of Engineers

Marisa C. Goulden, Tyndall Centre for Climate Change Research, University of East Anglia

Daniela Guitart, National Climate Change Adaptation Research Facility, Griffith University

John Handmer, Centre for Risk and Community Safety, RMIT University

Katharine Haynes, Risk Frontiers, Macquarie University

Sam S. L. Hettiarachchi, Department of Civil Engineering, University of Moratuwa

Saleemul Huq, International Institute for Environment and Development

Jiang Tong, China Meteorological Administration

David John Karoly, School of Earth Sciences, University of Melbourne

Jon E. Keeley, US Geological Survey, Western Ecological Research Center; Department of Ecology and Evolutionary Biology, University of California

Diane Keogh, Australian Centre for Sustainable Catchments, University of Southern Queensland

David King, Centre for Disaster Studies, James Cook University

Zbigniew W. Kundzewicz, Institute for Agricultural and Forest Environment, Polish Academy of Sciences, Poland; Potsdam Institute for Climate Impact Research

Timothy M. Kusky, Three Gorges Research Center for Geohazards, China University of Geosciences

Karine Laaidi, Institut de Veille Sanitaire – French National Institute for Public Health Surveillance

Alain Le Tertre, Institut de Veille Sanitaire – French National Institute for Public Health Surveillance

Gregor C. Leckebusch, School of Geography, Earth and Environmental Sciences, University of Birmingham

Matthew Mason, Risk Frontiers, Macquarie University

David M. Mills, Stratus Consulting Inc.

Helda Morales, El Colegio de la Frontera Sur, San Cristóbal de las Casas, Chiapas

Michael J. Mortimore, Drylands Research

Colette Mortreux, Department of Resource Management and Geography, University of Melbourne

Karen O'Brien, Department of Sociology and Human Geography, University of Oslo

Jean Palutikof, National Climate Change Adaptation Research Facility, Griffith University

Mathilde Pascal, Institut de Veille Sanitaire – French National Institute for Public Health Surveillance

Bimal K. Paul, Department of Geography, Kansas State University

Munshi K. Rahman, Kent State University

William D. Snook, City of Kansas City, Missouri, Health Department

Su Buda, China Meteorological Administration

Alexandra D. Syphard, Conservation Biology Institute

Melanie Thomas, Centre for Disaster Studies, James Cook University

Madeleine C. Thomson, International Research Institute for Climate and Society, and Mailman School of Public Health, Columbia University

Uwe Ulbrich, Institute of Meteorology, Freie Universitäet Berlin

Pier Vellinga, Earth System Science Department, Wageningen University; Research, Alterra, Wageningen

George Walker, Risk Frontiers, Macquarie University, Australia; Engineering and Physical Sciences, James Cook University; Aon Benfield Analytics

Joshua Whittaker, Centre for Risk and Community Safety, RMIT University

Preface

As we prepare this book for publication, communities here in Australia are emerging once again from a summer of disasters – widespread flooding, bushfires, and record heatwaves. Communities that have just recovered from the last summer of extremes (see Chapter 25) find their newly repaired roads, bridges, buildings, and homes once again destroyed or damaged. The question on the minds of many is whether this is the shape of things to come – is this the reality of climate change? But, this book is not about attribution of extreme events. It does not seek to tease out the signal of climate change from the background noise of day-to-day variability. Rather, its goal is to use the empirical evidence from extreme events to better understand our ability to respond effectively to climate change.

Many of us will experience climate change most clearly through a change in the frequency and severity of extremes. Although some acclimatisation will occur, how effectively we coped with extremes in the past can be indicative of how well we will adapt to climate change in the future. This book sets out to explore some of the lessons we can learn from past disasters and the adaptation that takes place after the event in preparation for the next. How successful are these actions, how do communities respond and behave, what makes an action fail, and how do these actions fare in the long term? These are some of the questions we wanted to address in this book in order to better understand how we might in the future adapt to climate change. As a modern society, we have no past experience of adapting to climate change, so that managing and responding to historical extremes provides the only empirical evidence, however imperfect, of how successfully we will adapt to climate change now and in the future.

The core of this book is a set of case studies of extremes in Australia, prepared for the National Climate Change Adaptation Research Facility, or NCCARF. These were some of the first products to come out of NCCARF, and we have found them to have resonance with communities and decision makers working in adaptation, giving them a focus for their own thinking about responding to climate change. From this core of Australian events we have built a set of case studies that seek to demonstrate how other countries, whether developed or developing, have approached the challenge.

Three of the editors – Sarah Boulter, Jean Palutikof, and Daniela Guitart – work in NCCARF, which is hosted by Griffith University in Queensland. The fourth editor is David John Karoly from the University of Melbourne. The original concept for the book arose from discussions between Jean and David, who worked together, respectively, as Head of Technical Support Unit and Coordinating Lead Author for Chapter 1 of the IPCC's Working Group II Fourth Assessment. The concept was then nurtured and fostered under the guidance of Sarah, and later Daniela helped to ensure that the idea became reality.

Sarah Boulter, NCCARF, Griffith University, Queensland
Jean Palutikof, NCCARF, Griffith University, Queensland
David John Karoly, University of Melbourne, Melbourne
Daniela Guitart, NCCARF, Griffith University, Queensland

1

Introduction

JEAN PALUTIKOF, SARAH BOULTER, DANIELA GUITART,
AND DAVID JOHN KAROLY

1.1 The Rationale for This Book

Every year extreme events occur which are costly in terms of human life and damage to infrastructure. Even in a stable climate, these events will occur as the tails of the normal distribution of events which characterise the climate of a place – climate variability. People and ecosystems will be partially adapted to such events but, especially in the case of severe events of great rarity, damage will ensue. The extent of this damage is a measure of the extent to which adaptation, whether planned or autonomous, has been successfully undertaken. Following the event, there will generally be an appraisal of the impact, the success or otherwise of the emergency management and the relative costs and benefits of additional adaptation measures. Where the conclusion is that the level of damage has been unacceptably high, it is likely that action will be taken to enhance coping capacity and resilience, especially when the costs are not prohibitive.

Extreme events, their impacts and the adaptations undertaken to protect ecosystems, human activities and human welfare contain instructive lessons for adaptation to climate change. First, understanding where, when and how action is taken to address vulnerabilities to extreme events is informative of the enabling factors needed to translate knowledge of climate change into adaptation. Second, understanding the reasons why some interventions designed to enhance adaptation to these extreme events succeed while others fail can help us to understand how to design effective adaptations to climate change. Finally, in many situations, the impacts of climate change will be felt principally through an increased frequency/severity of climate extremes. It is through adapting to this changed climatology of extremes that many adaptation actions to address climate change will take place.

This book explores climate extremes which have occurred in the recent past, in different geographical contexts, in the developed and developing world. By comparing the impacts, it seeks to identify what moves people to adapt, which adaptive activities succeed and which fail (and the underlying reasons for success or failure) and the

1

factors that determine when adaptation is indicated and when bearing the impact is a more appropriate response.

This book intends to inform about adaptation practice – what works, what does not, and why. Much has been written about the theory of adaptation, and high-level, especially international, policy responses to climate change. Although the literature on the practicalities of adaptation is, by comparison, remarkably small, there is a great need for such studies to inform our thinking. By presenting a set of case studies on adaptive responses to climate extremes, together with chapters which attempt to synthesise generally applicable conclusions from the outcomes of these case studies, this book begins to address this imbalance.

1.2 The Structure of This Book

The book is divided into six parts. The introductory part consists of this chapter and a discussion of disaster risk management and the interface with adaptation by Madeleine Thomson. The next four parts consist of case studies from North America, Australia, Europe and the Developing World respectively, and exploring extremes of droughts, floods, heatwaves and windstorms. These case studies examine the vulnerability and resilience of people, ecosystems and places prior to the event, disaster preparedness, emergency management responses during the event and responses after the event – the lessons learned, the barriers to adaptation and the characteristics of successful adaptation policies and activities. They use their findings to explore the implications for a world in which these events may become more frequent or intense.

Part I, covering North America, begins with a chapter by Andrew Garcia, looking at the impact of hurricane Katrina on New Orleans, and taking a primarily engineering perspective. It identifies the event as a case of clear forecasting success but social failure in responding. Sadly, Andrew passed away shortly after writing this chapter. We are deeply grateful to Kathleen White from the US Army Corps of Engineers for kindly helping us to proceed with the chapter. Chapter 4 is an examination of the history of flooding in the Mississippi River Basin by Timothy Kusky, leading to a discussion on the effectiveness of flood control measures in constricted rivers and urbanised floodplains. Jon Keeley, Alexandra Syphard and C.J. Fotheringham in Chapter 5 examine wildfires in Southern California and the vulnerability of human settlements located in and around fire-prone wildland vegetation. David Mills and William Snook close out Part I by exploring, in Chapter 6, the evolution of the extreme heat program established in Kansas City, Missouri, in 1980 in response to a severe heatwave in that year, and the lessons to be learned from the revision and accrued experience of that program.

Part II contains case studies from Australia, a country where the climate has been described as 'boom or bust'. In Chapter 7, Linda Botterill and Stephen Dovers examine the process of drought policy change and issues of water allocation in the

Murray-Darling Basin under conditions of water scarcity and climate variability. It explores the move from disaster-focused responsive policy-making to long-term planning and building resilience. In Chapter 8, Joshua Whittaker, John Handmer and David Karoly explore the environmental context and human dimensions of the 2009 'Black Saturday' bushfire disaster in south-east Australia. They outline the findings of the Royal Commission set up to enquire into the fires, the resulting policy and management changes and the community response. In Chapter 9, Matthew Mason, Katharine Haynes and George Walker look at the steps taken to revise building regulations and construction practices following cyclone Tracy, which struck Darwin in 1974. With 60 per cent of homes destroyed by the cyclone, the goal was to ensure that such levels of damage could never be repeated in Australia, and this chapter considers the long-term success of the ensuing changes to building regulations. David King, Armando Apan, Diane Keogh and Melanie Thomas in Chapter 10 use questionnaire surveys to understand how communities and businesses in a small inland outback town and a medium-sized coastal city of Queensland coped with the 2008 floods, and what measures they took to lessen the impact of future events.

Part III looks at case studies of windstorm, heatwave and flood events in Europe. Uwe Ulbrich, Gregor Leckebusch and Markus Donat begin in Chapter 11 by reviewing windstorms as one of the most common and most costly natural disasters affecting Europe. They note that part of the required knowledge of windstorm risk is based on proprietary risk models, available to insurance brokers on a commercial basis from companies specialising in these services. This market-driven 'vulnerability' has potentially negative implications for timely adaptation. In Chapter 12, Mathilde Pascal, Alain Le Tertre and Karine Laaidi take a public health perspective on adaptation to heatwaves, drawing on the lessons learned in France from the 2003 heatwave. They highlight the need for both long-term adaptation through changing behaviours, climate-sensitive housing design and reducing urban heat islands in cities, and immediate responses based on heat warning systems and government-developed and legislatively-enforced heat prevention plans. The two chapters that follow discuss lessons learned about flood management based, respectively, on the 1997, 2002 and 2010 floods in Central Europe (Zbigniew Kundzewicz, Chapter 13) and on the 1953 North Sea storm in the Netherlands (Pier Vellinga and Jeroen Aerts, Chapter 14). The first study highlights socio-economic factors leading to increased vulnerability including development, urbanisation and political transition. The latter provides an example of strong political and community will to prevent the repeat of a catastrophic disaster through large-scale economic investment supported by pragmatic cost-benefit analysis and the adoption of a future-forward acceptable risk baseline.

Part IV looks at case studies of extreme events in the developing countries of Africa, Asia and Central America. In Chapter 15, Simon Batterbury and Michael Mortimore explore adaptive measures taken by Sahelian people to minimise the risk of lost food production during severe droughts, and argue that social and community-driven change represents these communities' successful adaptation to changing climate conditions. In Chapter 16, the only one on a non-climatic event, but still informative for thinking around climate change adaptation, Sam Hettiarachchi and Priyan Dias explore the effects of the 2004 Indian Ocean tsunami on Sri Lanka. They highlight the importance of geographically distributing assets and infrastructure, and the need for combined approaches (e.g. early warning systems together with engineered protection) to reduce vulnerability. In Chapter 17, Bimal Paul and Munshi Rahman look at recovery from the impacts of the 2007 cyclone Sidr in Bangladesh, focusing specifically on progress made to date on housing reconstruction. The case study highlights the challenges of disaster recovery in developing nations and explores the concept of a 'build back better' approach to aid recovery and address development needs. Hallie Eakin, Helda Morales, Edwin Castellanos, Gustavo Cruz-Bello and Juan Barrera (Chapter 18) take a socio-ecological approach to exploring adaptation to tropical storms by coffee producers in Guatemala and Mexico. They highlight the importance of empowering local communities to act on their existing knowledge and experience, and of building on local social institutions to enhance local risk management capacity. In Chapter 19, Marisa Goulden and Declan Conway look at responses to the 1997–1998 floods in the Upper White Nile, linking strategies for disaster risk reduction and climate change adaptation. The concluding chapter of this part, by Zbigniew Kundzewicz, Jiang Tong and Su Buda, discusses both structural and non-structural (e.g. forecasting and modelling) flood control measures taken following several floods in the Yangtze River, China, and a growing awareness of increasing flood risk.

Together, these case studies explore in depth our knowledge of present-day vulnerabilities, adaptation and resilience to climate variability. In the concluding Part V, the lessons from these case studies are drawn together in a set of synthesising chapters to address the overarching goal of informing our thinking about adaptation to future climate change. In Chapter 21, Jessica Ayers, Saleemul Huq and Sarah Boulter examine the interrelationships between development, vulnerability and adaptive capacity, considering how these have influenced disaster responses in developed and developing countries. They highlight that, even though there is strong evidence that low development frequently determines vulnerability to hazards, it also creates entry points for development-based and participatory disaster responses that are often missed in high-income contexts. This is followed by Karen O'Brien and Thomas Downing in Chapter 22 looking at how climate change has affected, and will affect, vulnerabilities to disasters and disaster risk management. They argue that climate change is a 'game changer' that requires new ways of thinking about managing extreme events. They conclude that climate change calls for more than taking into account the lessons learned from vulnerabilities and responses to past events and applying these to future scenarios; rather, it calls for transformative approaches to reducing risk in a changing climate. In Chapter 23, Jon Barnett, Colette Mortreux and Neil Adger draw insights from the case studies to illuminate their thinking about the limits and barriers to climate change adaptation, and the risk of maladaptation. In Chapter 24, Sarah Boulter, Jean Palutikof and David Karoly consider the commonalities and points of difference among the case studies and synthesising chapters to consider what lessons emerge from the book for climate change adaptation.

The book is completed by an afterword on what happens when multiple natural disasters affect the same location in a short period of time (David Karoly and Sarah Boulter, Chapter 25). It takes the case of the Australian experience in 2010–2011, when wildfires, heatwaves, flooding, cyclones and insect plagues challenged almost the entire continent across a summer period. This multiplicity of events may be one of the greatest risks we face from future climate change. At the present time, such occurrences are rare, yet when they do occur they offer one of the few opportunities we have to empirically explore what it may be like to live with climate change.

2

Climate Change and Disaster Risk Management: Challenges and Opportunities

MADELEINE C. THOMSON

An increase in extreme weather and climate events (e.g. floods, droughts, heatwaves, windstorms, etc.) is widely understood as one of the expected outcomes of global climate change (IPCC, 2012). Over the last forty years the disaster response community have increasingly focused on prevention and preparedness for such natural (hydrometeorological) disasters, moving from disaster response to disaster risk management and towards disaster risk reduction (UNISDR, 2011), and are increasingly engaged in the discourse on climate change adaptation (see Figure 2.1).

Recent developments in climate science, climate services, geo-referenced data assimilation, management, analysis and dissemination tools and communication technologies have dramatically improved the opportunity for increasing resilience of communities to weather- and climate-related disasters. Here the challenges posed by gaps in our understanding of climate change and disaster risk in terms of space and time scales, evidence-based action and regulations and policies are discussed alongside new knowledge, technologies and policy frameworks that may help to overcome these gaps and make a difference on the ground.

2.1 Challenges of Climate Change and Disaster Risk: Space and Time Scales

Attributing disasters associated with extreme weather events to anthropogenic changes in the climate system remains a major challenge, although there is increasing interest in making progress in this area. It has commonly been stated that no individual extreme event can be attributed to climate change, but increasingly the probability that global climate change has influenced individual events (or not) is a subject for scientific discussion (Peterson et al., 2012). In the language of the Intergovernmental Panel on Climate Change (IPCC, 2012), 'it is likely that anthropogenic influences have led to warming of extreme daily minimum and maximum temperatures at the global scale' and that 'there is medium confidence that anthropogenic influences have contributed to intensification of extreme precipitation at the global scale'.

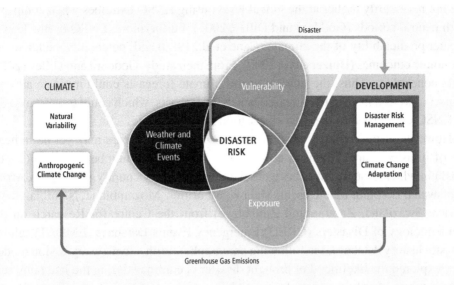

Figure 2.1 Illustration of the core concepts 'Managing the Risks of Extreme Events and Disasters to Advance Climate Change Adaptation' (IPCC, 2012, figure SPM1). Human-induced climate change and natural climate variability interacts with socio-economic development (via vulnerability and exposure) and the responses from the disaster risk management and climate change adaptation communities to influence disaster risk.

A key challenge is the fact that climate change is best observed at the global and regional scales but population vulnerabilities and many significant climate/weather phenomena occur at sub-regional, national or sub-national scales. Thus the disaster risk community is faced with a need to reconcile the interaction of a global-scale phenomenon with local-scale challenges. We illustrate this issue of scale with the example of the global and local impact of the El Niño–Southern Oscillation (ENSO) phenomenon.

Much of the evidence for climate change occurring over century-long time scales comes from large-scale increases in upper-ocean temperatures that are evident in observational records (Gleckler et al., 2012). At shorter (one year or two) time scales, the most significant natural global climatic driver is the ENSO phenomenon. This is a periodic warming or cooling of sea surface temperatures (SSTs) in the eastern and central equatorial Pacific, which generates a significant proportion of short-term climate variations. Since ENSO (El Niño for warming SSTs and La Niña for cooling SSTs) and its climate impacts were first documented in the 1980s (Ropelewski and Halpert, 1987), the prevailing assumption has been that more severe and widespread climate anomalies – and, therefore, greater climate-related socioeconomic losses – should be expected during ENSO extremes (Gueri et al., 1986; Bouma, 1997). Contrary to expectations, a global-scale study indicated that disaster losses are not necessarily higher at the global level during ENSO extremes when compared with neutral periods (Goddard and Dilley, 2005). Furthermore, ENSO events lead to greater predictability of the climate (Cane et al., 1986) and, potentially, better socio-economic outcomes (Buizer et al., 2000). From their study, Goddard and Dilley (2005: 651) concluded, 'Thus, the prudent use of climate forecasts could mitigate adverse impacts and lead instead to increased beneficial impacts, which could transform years of ENSO extremes into the least costly to life and property'.

However, at the regional or sub-regional scale, ENSO years may yet be indicative of disaster years. Thomson et al. (2003) obtained the number of years (1980–2001) in which drought disasters were recorded for every country in southern Africa (Botswana, Lesotho, Madagascar, Malawi, Mauritius, Mozambique, Namibia, South Africa, Swaziland, Zambia and Zimbabwe) from the Centre for Research on the Epidemiology of Disasters (CRED) Emergency Events Database (EMDAT) (global disaster history database; see following paragraphs). With a logistic regression model, they explored the likelihood of drought disaster occurrence during the late rainy season (January–March) the year after an El Niño event when ENSO impacts are known to occur in Southern Africa. They recorded a 120 per cent increase in probability of drought disaster in the year after El Niño onset ($p = 0.0005$) when compared with other times.

Given the difficulties in attribution and the uncertainties in predicting changes in future risks at regional or local scales, the disaster risk management community must take a highly pragmatic approach to reducing disaster risk associated with hydro-meteorological events in a changing climate.

To date the most widely used global resource that provides a routinely updated assessment of the impact of natural disasters on societies around the world comes from CRED,[1] based at the Catholic University of Louvain, Brussels. CRED manages a routinely updated disaster database which records disaster losses in terms of economic costs, number of deaths and numbers of persons affected, by type of disaster. The CRED disaster database has been instrumental in providing data to support the current consensus that the economic impacts of hydro-meteorological disasters are increasing (IPCC, 2012). What is not so clear, however, is whether the higher impacts stem from an increase in weather and climate extremes or an increase in exposure, as populations continue to rise and economic growth inflates figures for economic damage during any given disaster (UNISDR, 2011). Noteworthy is that a high proportion of disasters in the CRED database (and other more detailed national databases such as DesInventar,[2] developed for Latin American countries) are not necessarily associated with extreme events per se but rather with the interaction between climate and environmental changes and high levels of societal vulnerability. Thankfully the CRED database also

---

[1] http://www.cred.be/   [2] http://online.desinventar.org/desinventar/

indicates that the loss of life per disaster is decreasing – the fact which may be attributed, in part, to improved early warning and response systems (Golnaraghi, 2012).

Without indicators that can correctly attribute the drivers of disaster increases, the past is not a particularly reliable guide to understanding potential losses (i.e. economic, lives) from hydro-meteorological disasters in the future. This is particularly so because climate change may have tertiary impacts on vulnerability such as, for example, disruption of global or local economies or food supply systems and the risk of infectious diseases. Additionally, the relationship of vulnerability and the impact of hydro-meteorological events may be highly non-linear. Given the uncertainty, a priority must be given to the development of a 'climate-smart' disaster community – one that is conversant with a basic knowledge of the climate system and an understanding of how the climate's natural variability on multiple time scales interacts with global environmental change and societal vulnerabilities to produce devastating societal impacts. It is particularly important to focus on the utility of information coming from the climate community in solving practical problems. Short-term weather forecasts for extreme events are proving astonishingly accurate when available. However, in many developing countries weather forecasts rarely extend beyond one to two days, and access to the information by the population at large is often limited. Seasonal climate forecasts are still underutilised, and little is known about the potential predictability of extreme events, although there are indications that such intra-seasonal information may become available in some regions (Robertson et al., 2009). Many decision makers would like to incorporate climate change information into their decision process, but the century-long span of typical climate change projections does not fit with their operational outlook that often spans only a few years. For those planning infrastructure developments the operational focus may extend to a few decades, but demand for climate information at the decadal time scale may well be running ahead of supply. This is because, while decadal predictions are high on the climate science research agenda (Meehl et al., 2011), there is considerable uncertainty as to whether the practical requirements for this type of information can be met anytime soon – if at all (Cane, 2010).

While it is increasingly understood that effective early warning must explicitly address the spatio-temporal dimensions and estimations of uncertainty of the climate and weather drivers of disasters, to date this knowledge is rarely translated into practical methodologies for operational use. The 'translation of the best science on climate change into practical action' is now an urgent priority (M. Wahlström,[3] personal communication, 2011).

---

[3] UN Secretary-General's special representative for disaster reduction and head of the UN International Strategy for Disaster Reduction (UNISDR) when speaking at the 'high-level dialogue and adaptation roundtable: global sustainability in a changing climate'.

For instance, over the last decade the Horn of Africa has experienced an increasing frequency of drought events, particularly during the 'long rains' which typically runs from March to May (Lyon and DeWitt, 2012). The most recent of these events, in 2010–2011, was associated with La Niña and considered the most severe in sixty years (Williams and Funk, 2011). Of importance is the fact that the observed drying differs from climate change projections, which suggest that the climate of East Africa will become wetter by the end of the current century (IPCC, 2007). The apparent discrepancy between recent drying and projections for a wetter future raises some fundamental questions to climate scientists and poses particular problems for regional policymakers who must make strategic decisions for many millions of people with conflicting information. Should they prepare societies for floods or droughts? Furthermore, if climate change scenarios are of too long a time scale for practical use, decadal climate forecasts are still on the research bench, and seasonal forecasts still underexplored, what should practitioners expect from the climate community to help to manage climate-related disaster risks?

2.2 Knowledge for Action

Following the devastating Sahelian droughts of the 1970s and 1980s, the disaster response community sought to prioritise prevention over more traditional response capacities and to identify disaster risk management (including preparedness and early warning systems) as key to economic and social development (Blench and Marriage, 1999). It soon became clear that technological capabilities for famine early warning alone cannot guarantee a timely response, either by international donors or national governments. It is critical to know who 'owns' early warning information and how the information is used in order to understand whether or not the information could result in an appropriate and timely response (Buchanan-Smith and Davies, 1995). In recent years, early warning systems for natural disasters that are tied to response capability have emerged as a priority for the disaster and development communities (IFRC, 2009). The incorporation of climate change into a disaster response portfolio of risk management has further reinforced the need for better data, methodologies and tools as well as policies and practices. In particular, the potential for climate change to increase the likelihood of extreme events has resulted in a renewed focus on early warning for adaptation. The idea is that better preparation for climatic shocks today will build capacities to better manage an increasingly extreme climate in the future (Hellmuth et al., 2011).

The recent occurrence of hurricane/tropical storm Sandy, which hammered the eastern seaboard of the United States in October 2012, illustrates the impact of an extreme weather disaster in a highly developed urban environment. The hurricane forecasts provided by global centres, particularly the European Centre for Medium Range Weather Forecasting (Hewson, 2012), proved highly informative and were acted upon by local and national officials days ahead of the disaster. New York City Mayor Michael Bloomberg closed the city's financial district and ordered a mandatory evacuation of low-lying areas as a result of predictions indicating an unprecedented surge in sea level, undoubtedly saving many lives in the process. The subsequent presidential request to Congress for a US$60.4 billion fund to rebuild the superstorm-ravaged region provides an indication of the extent of economic losses (Russ and Lawder, 2012), while the initial observation that Sandy had minimum impact on the US national economy provides evidence of the level of resilience of a developed society (PNC, 2012). The infrastructure devastation caused by Sandy had been largely predicted in a 2011 report 'Responding to Climate Change in New York State' by the New York State Energy Research and Development Authority (Rosenzweig et al., 2011), and the disaster reignited climate change as a political issue in the US 2012 elections.

The contrast of preparedness and response for the impact of hydro-meteorological disasters in poor regions of the world is stark. In the recent drought in the Horn of Africa, alarming levels of acute malnutrition were documented beginning in March 2010, and by August 2010, an impending food crisis was forecasted following the failure of the rains. Despite these observations, the situation remained unrecognised, and further deterioration in the health of the affected populations was entirely predictable and made worse by the failure of the rains in 2011. By the time famine was officially declared by the United Nations (20 July 2011), levels of malnutrition and mortality exceeded catastrophic levels (Kim and Guha-Sapir, 2012). It was only at this stage in the crisis that the full response mechanisms of the global humanitarian community were finally triggered (Hillier and Dempsey, 2012). In theory, population vulnerability along with the spatial and temporal scales of a disaster should determine the scale of the appropriate response, which in turn determines nearly every other aspect of the disaster risk management process, among them: Which actors must be involved? In which capacity? How much preparation is feasible and desirable? How skilful does a prediction of a disaster need to be to elicit action? It seems that while new technologies are in place, major gaps remain in policy and practice that are needed to ensure the timeliness of appropriate responses.

Within the disaster community, infectious disease epidemics are identified as both an outcome of natural disasters (most commonly floods) and as a natural disaster in their own right. There is now a fairly extensive literature on the relationship of epidemics of climate-sensitive diseases to ENSO phenomena (e.g. Kelly-Hope and Thomson, 2008), and the use of climate information in epidemic prediction has been an integral part of the World Health Organization (WHO) strategy for the control of epidemic malaria for the last decade (WHO, 2001). However, even with good evidence on the likelihood of unusual increases in malaria cases as a result of a forecasted heavy rainfall season (Thomson et al., 2005; Thomson et al., 2006), there are still many challenges in ensuring an effective response to epidemic early warnings. One particular limitation of seasonal climate forecasts is their level of uncertainty which varies from region to region, season to season, and year to year (Barnston et al., 2010). The WHO framework for malaria early warning systems embeds the use of seasonal climate forecasts information in a climate risk management strategy that incorporates a thorough understanding of the historical climate-related risks and population vulnerability to epidemics, up-to-date information on the current climate, as well as predictions of the future climate when and where they are considered reliable and useful. An example of the framework's implementation occurred following the 2011 famine crisis in the Horn of Africa. WHO requested risk information from the International Research Institute for Climate and Society for malaria epidemics in the burgeoning refugee camps for Somalis fleeing the 2010–2011 drought-induced famine. The seasonal climate forecast for the 2011 short rains from the Greater Horn of Africa Climate Outlook Forum (CHACOF) was considered as one source of potentially valuable information.[4] However, given its high uncertainty, the forecast information available was also supported by historical information on the climate suitability for malaria transmission in the region (Grover-Kopec et al., 2006) created using a high-quality national database on climate and temperature (Dinku et al., 2011). Based on the possible risk (though low probability) of epidemics in the highly vulnerable camp populations, WHO and its partners agreed to provide an additional 5 million doses of the highly effective anti-malaria drugs (Artemesin-based Combination Therapies) to the region (A. Bosman, WHO Collaborating Centre, personal communication, December 2012).

While impoverished populations in developing countries are most at risk of health impacts of hydro-meteorological disasters, hurricane Katrina in the United States, as a result of which more than 3,000 people died (Garcia, Chapter 3 in this volume), and the 2003 heatwaves in Europe, which resulted in more than 70,000 deaths (Pascal et al., Chapter 12 in this volume; Robine et al., 2008), demonstrate that vulnerable populations – particularly the poor and elderly – in the developed world may also be severely affected. Changing demographics in developed countries are increasing the proportion of vulnerable elderly and isolated people in the population, and this must be factored into the development of early warning and response systems (Kovats and Ebi, 2006) (see the example of Kansas City in Chapter 6 in this volume).

Building capacity in the disaster response community to use climate and environmental information in disaster risk assessment, risk reduction and disaster response requires a long-term strategy. Practitioners use what they learn in their routine training, respond to peer perceptions and institutionalised processes, and are much more confident in using new information if they perceive that the evidence base for its use has been properly established. When introducing a new technology (such as seasonal climate forecasts), it is important to consider how it is introduced, what community

---

[4] Statement from the twenty-ninth greater Horn of Africa Climate Outlook Forum (GHACOF 29), 1–3 September 2011, Entebbe, Uganda.

capacities need to be built and how the new knowledge gained can be turned into improved decision making (Braman et al., 2013). The uptake and use of seasonal climate forecasts is likely to be easier in a community that already routinely uses meteorological data and remote-sensing products in decision making – such as, for example, the food security community. Lack of access to high-quality meteorological data needed to develop the evidence of the impact of climate anomalies on disasters is a major challenge to the use of climate information; in the absence of quality data, the existing data may be used inappropriately or even ignored altogether (Thomson et al., 2011).

The 'Health Risk Management in a Changing Climate' project, funded by the Rockefeller Foundation and executed by the Red Cross/Red Crescent Climate Centre, was an attempt to build globally relevant operational knowledge on how to address changing risk by focusing on extreme events in four developing countries. In Tanzania and Kenya there was a particular focus on diarrheal and vector-borne diseases, and in Vietnam and Indonesia there was a particular focus on dengue fever. The project underscored the need for communicating scientific knowledge of climate change in a practical way and providing substantial guidance on how to deal with the different concepts and perspectives of the climate and disaster community.

2.3 Changing Policies and Practices That Can Make a Difference on the Ground

The Hyogo Declaration and Framework for Action (HFA) was a key outcome of the World Conference on Disaster Reduction held in Kobe, Hyogo, Japan, in early 2005 (UNISDR, 2007), following the catastrophic tsunami affecting coastal communities in Sri Lanka (Hettiarachchi and Dias, Chapter 16 in this volume), India, Thailand and Indonesia. The HFA was the first plan to explain, describe and detail the work that is required from all different sectors and actors to reduce disaster losses. As part of this process, it has focused on the use of databases for disaster risk reduction, having observed that much of the existing operational research related to emergencies and disasters lacks consistency, is of poor reliability and validity and is of limited use for establishing baselines, defining standards, making comparisons or tracking trends (V. Murray, Health Protection Agency, personal communication, 2011).

Subsequently, concerns raised about the implications of climate change for disaster risk management led to the agreement between the IPCC and the United Nations International Strategy for Disaster Reduction (UNISDR) to undertake a Special Report (SREX) on 'Managing the Risks of Extreme Events and Disasters to Advance Climate Change Adaptation' (IPCC, 2012). At last it seems that the necessary institutional links are being made across institutions, communities and time scales.

Since Kobe, the idea of a single national platform for disaster risk early warning (Multi-Hazard Warning System) has gained ground (Golnaraghi, 2012). The rationale advanced for housing all the rapid onset disaster early warning systems (including, as appropriate, seismological, coastal and marine hazards) in a common platform based at the national meteorological agency is because these organisations already operate 24/7 and routinely provide emergency information for extreme weather events (such as hurricanes and heatwaves). Disasters triggered by anomalies in the hydrological cycle are the most frequent causes of disasters worldwide. For example, epidemics, floods, droughts and windstorms account for more than 80 per cent of the disasters that afflict Africa (UNISDR, 2003), where they already pose a major challenge to sustainable development. The risk associated with floods, droughts and extreme weather is directly related to the characteristics (intensity and extent) of the hazard and the characteristics (vulnerability) of the population. This is clearly stated in the SREX IPCC report to policymakers (IPCC, 2012) where evidence is presented to support the following statements with regard to disaster losses:

1. Economic losses from weather- and climate-related disasters have increased, but with large spatial and inter-annual variability.
2. Economic, including insured, disaster losses associated with weather, climate and geophysical events are higher in developed countries. Fatality rates and economic losses expressed as a proportion of gross domestic product (GDP) are higher in developing countries.
3. Increasing exposure of people and economic assets has been the major cause of long-term increases in economic losses from weather- and climate-related disasters (high confidence). Long-term trends in economic disaster losses adjusted for wealth and population increases have not been attributed to climate change, but a role for climate change has not been excluded.

The development of a Global Framework for Climate Services (GFCS) sits at the heart of this emerging consensus for an integrated approach (Hewitt et al., 2012). The GFCS is now an overarching strategic priority for the World Meteorological Organization, which guides all of its work with national meteorological agencies. It therefore provides an important opportunity for the disaster community to establish a comprehensive approach to better understand impacts and support disaster risk reduction through better management of climate risks.

Climate services differ from traditional weather services in that they consist of information on changes in the statistical probability of the climate over extended periods – such as the likelihood of a forthcoming season being unusually wet or dry as indicated through a seasonal climate forecast. When combined with other sources of information, climate services may extend to include information on likely impacts such as agricultural drought, floods, landslides and epidemics.

Central to the Framework is a User Interface Platform (UIP), which provides a structured means for users, user representatives, climate researchers and climate service providers to interface at global, regional, national and community levels. The platform aims to achieve four outcomes in feedback, dialogue, evaluation and literacy. To be successful, it must be built in a way that is responsive to the expressed needs of the intended user base – and yet, the user base is often naïve as to (1) what can be delivered and (2) how the services that might be delivered can improve outcomes that they care about. The challenge is particularly profound in developing countries where climate variability and change have long been understood as a major social and economic threat, yet the use of weather and climate information to mitigate that threat in practice is very limited. The International Research Institute for Climate and Society (IRI) Gap Analysis for the Global Climate Observing System in Africa (IRI, 2006) expressed this gap between perceived and realised value as due to 'Market Atrophy'. This is where potential users have no knowledge or experience in what could be provided and therefore fail to ask for relevant services while the supplier (in this case the National Meteorological Agency), in the absence of articulated demands, fail to produce relevant services. Without relevant information from the climate community the potential user either gives up or goes elsewhere.

In Africa, as an example, important areas of improvement in recent years include overcoming key challenges identified in the Gap Analysis (IRI, 2006) in policies, practices, services and data, through a focus on better acquisition, management and sharing of observational data, improved service provision from the meteorological agencies to potential end users, capacity building in relevant practitioner communities to use climate information, and improving the policy environment that enables better decision making based on evidence.

Current policies for disaster response often rely on post-disaster information (e.g. numbers of people affected, proportion of children malnourished) to initiate the release of emergency funds (as observed in the Horn of Africa 2010–2011 response). An important change needed for improving the response options is that disaster relief donors need to be prepared to fund prevention activities based on warnings of potential disasters even when the potential societal impacts have a high degree of uncertainty.

An example of such a shift in policy happened in 2008 in West Africa, when the International Federation of Red Cross/Red Crescent Societies (IFRC) issued an appeal for supplies weeks before any event occurred. It based the appeal on seasonal climate forecasts, issued in May 2008, which indicated a heightened chance of above-normal rainfall during West Africa's July-to-September rainy season. Concerned about climate change, and having been caught off guard by devastating floods in West Africa the year before, the IFRC and their national societies in the region were eager to respond early and, on the basis of the forecast successfully launched an emergency appeal (Braman et al., 2013).

2.4 Conclusion

While the challenges of climate change are many for the disaster risk management community, so are the opportunities. Because of the sense of urgency that something different must be done, climate change represents a unique opportunity to develop new strategies and new ways of thinking about disaster risk management as a whole;

included in this new way of thinking is access and effective use of climate knowledge and information. As part of this strategic shift there must be a sea change in the way operational research related to disaster response is undertaken, especially as it refers to disaster-related morbidity and mortality. Access to timely, relevant and quality-assured data has been identified as key. For instance, without accurate high spatial and temporal resolution and historical and current climate data, it is impossible to undertake detailed impact analysis that can be used to create the type of evidence policymakers need. While valuable data may be collected at the individual-project level (Hellmuth et al., 2011), there is rarely sufficient real-time evidence on whether – and if so, how – outcomes are improving or deteriorating because of the management decisions and actions taken (Sheikh and Musani, 2006). Despite the increasing interest and implementation in disaster risk reduction through early warning of natural hazards and extreme weather events, there is limited evidence of the cost-effectiveness of such warning systems. This is the case even in developed countries where, for example, a number of operational early warning systems for heat waves have been deployed in recent years (Lowe et al., 2011).

New initiatives seeking to strengthen the evidence base for how best to address the societal impacts of natural and humanitarian disasters are a step in the right direction. It is hoped that the case studies in this book will also provide the type of evidence needed to improve policies and practical decisions when disaster strikes.

Acknowledgements

With thanks to Patrick Robbins for his assistance with the early phase of manuscript development, Maarten Van Aalst for helpful discussions and Simon Mason and Gilma Mantilla for comments on an earlier draft of the manuscript.

References

Barnston, A. G., Li, S., Mason, S. J. et al. (2010). Verification of the first 11 years of IRI's seasonal climate forecasts. *Journal of Applied Meteorology and Climatology*, 49, 493–520.

Blench, R. and Marriage, Z. (1999). *Drought and Livestock in Semi-arid Africa and Southwest Asia. Working Paper 117*. London, UK: Overseas Development Institute.

Bouma, M. J., Kovats, R. S., Goubet, S. A., Cox, J. S. and Haines, A. (1997). Global assessment of El Niño's disaster burden. *Lancet*, 350, 1435–1438.

Braman, L., van Aalst, M. K., Mason, S. J. et al. (2013). Climate forecasts in disaster management: Red Cross flood operations in West Africa, 2008. *Disasters*, 37 (1), 144–164.

Buchanan-Smith, M. and Davies, S. (1995). *Famine Early Warning and Response: The Missing Link*. London, UK: Intermediate Technology Publications.

Buizer, J. L., Foster, J. and Lund, D. (2000). Global impacts and regional actions: preparing for the 1997–98 El Nino. *Bulletin of the American Meteorological Society*, 81, 2121–2139.

Cane, M. (2010). Climate science: decadal predictions in demand. *Nature Geoscience*, 3, 231–232.

Cane, M. A., Zebiak, S. E. and Dolan, S. C. (1986). Experimental forecasts of El Niño. *Nature*, 321, 827–832.

Dinku, T., Hilemariam, K., Grimes, D., Kidane, A. and Connor, S. (2011). Improving availability, access and use of climate information. *World Meteorological Bulletin*, 60 (2), 80–86.

Gleckler, P. J., Santer, B. D., Domingues, C. M. et al. (2012). Human-induced global ocean warming on multidecadal timescales. *Nature Climate Change*, 2, 524–529.

Goddard, L. and Dilley, M. (2005). El Nino: catastrophe or opportunity. *Journal of Climate*, 18, 651–665.

Golnaraghi, M. E. (2012). *Institutional Partnership in Multi-Hazard Early Warning Systems. A Compilation of Seven National Good Practices and Guiding Principles*. Dordrecht: Springer.

Grover-Kopec, E., Blumenthal, B., Ceccato, P. et al. (2006). Web-based climate information resources for malaria control in Africa. *Malaria Journal*, 5, 38.

Gueri, M., Gonzalez, C. and Morin, V. (1986). The effect of the floods caused by El Niño on health. *Disasters*, 10, 118–124.

Hellmuth, M. E., Mason, S., Vaughan, C., van Aalst, M. and Choularton, R. (eds.) (2011). *A Better Climate for Disaster Risk Management*. New York: International Research Institute for Climate and Society.

Hewitt, C., Mason, S. and Walland, D. (2012). The global framework for climate services. *Nature Climate Change*, 2, 831–832.

Hewson, T. (2012). ECMWF forecasts of 'Superstorm Sandy'. *ECMWF Newsletter*, 133, 9–11.

Hillier, D. and Dempsey, B. (2012). *A Dangerous Delay: The Cost of Late Response to Early Warnings in the 2011 Drought in the Horn of Africa*. Oxford, UK: Oxfam and Save the Children. Accessed 25 January 2013 from: <http://www.oxfam.org/en/policy/dangerous-delay>.

IFRC – International Federation of Red Cross (2009). *World Disasters Report 2009: Focus on Early Warning Early Action*. Geneva, Switzerland: International Federation of Red Cross and Red Crescent Societies.

IPCC – Intergovernmental Panel on Climate Change (2007). *Climate Change 2007: Synthesis Report. Contribution of Working Groups I, II and III to the Fourth Assessment Report of the Intergovernmental Panel on Climate Change*, eds. Core Writing Team, R. K. Pachauri and A. Reisinger. Cambridge, UK: Cambridge University Press.

IPCC – Intergovernmental Panel on Climate Change (2012). Summary for policymakers. In *Managing the Risks of Extreme Events and Disasters to Advance Climate Change Adaptation. A Special Report of Working Groups I and II of the Intergovernmental Panel on Climate Change*, eds. C. B. Field, V. Barros, T. F. Stocker et al. Cambridge, UK: Cambridge University Press, pp. 3–21.

IRI – International Research Institute for Climate and Society (2006). *A Gap Analysis for the Implementation of the Global Climate Observing System Programme in Africa*. New York: International Research Institute for Climate and Society.

Kelly-Hope, L. A. and Thomson, M. C. (2008). Climate and infectious disease. In *Seasonal Forecasts, Climatic Change, and Human Health*, eds. M. C. Thomson, R. Garcia-Herrera and M. Beniston. Dordrecht: Springer, pp. 31–70.

Kim, J. J. and Guha-Sapir, D. (2012). Famines in Africa: is early warning early enough? *Global Health Action*, 5, 18481.

Kovats, R. S. and Ebi, K. L. (2006). Heatwaves and public health in Europe. *European Journal of Public Health*, 16, 592–599.

Lowe, D., Ebi, K. L. and Forsberg, B. (2011). Heatwave early warning systems and adaptation advice to reduce human health consequences of heatwaves. *International Journal of Environmental Research and Public Health*, 8, 4623–4648.

Lyon, B. and DeWitt, D. G. (2012). A recent and abrupt decline in the East African long rains. *Geophysical Research Letters*, 39 (2), L02702.

Meehl, G. A., Goddard, L., Murphy, J. et al. (2011). Decadal prediction – can it be skillful? *Bulletin of the American Meteorological Society*, 90, 1467–1485.

Peterson, T. C., Stott, P. and Herring, S. (2012). Explaining extreme events of 2011 from a climate perspective. *Bulletin of the American Meteorological Society*, 93, 1041–1067.

PNC (2012). *Executive Summary: Recovery will continue after Election and Superstorm Sandy. November/December National Economic Outlook*. Pittsburgh, PA: PNC.

Robertson, A. W., Moron, V. and Swarinoto, Y. (2009). Seasonal predictability of daily rainfall statistics over Indramayu district, Indonesia. *International Journal of Climatology*, 29, 1449–1462.

Robine, J.-M., Cheung, S. K., Roy, S. et al. (2008). Death toll exceeded 70,000 in Europe during the summer of 2003. *Comptes Rendus Biologies*, 331, 171–178.

Ropelewski, C. F. and Halpert, M. S. (1987). Global and regional scale precipitation patterns associated with the El Niño/Southern Oscillation. *Monthly Weather Review*, 115, 1606–1626.

Rosenzweig, C., Solecki, W., DeGaetano, A. et al. (2011). *Responding to Climate Change in New York State: The ClimAID Integrated Assessment for Effective Climate Change Adaptation. Synthesis Report*. New York: New York State Energy Research and Development Authority.

Russ, H. and Lawder, D. (2012). Obama seeks $60.4 billion for Sandy repairs, states want more, *Reuters*, 7 December 2012. Accessed 25 January 2013 from: <http://www.reuters.com/article/2012/12/08/us-usa-stormsandy-aids-idUSBRE8B61AH20121208>.

Sheikh, A. and Musani, A. (2006). Emergency preparedness and humanitarian action: the research deficit. Eastern Mediterranean Region perspective. *Eastern Mediterranean Health Journal*, 12 (2), 54–63.

Thomson, M. C., Abayomi, K., Barnston, A. G., Levy M. and Dilley, M. (2003). El Niño and drought in southern Africa. *Lancet*, 361 (9355), 437–438.

Thomson, M. C., Doblas-Reyes, F. J., Mason, S. J. et al. (2006). Malaria early warnings based on seasonal climate forecasts from multi-model ensembles. *Nature*, 439, 576–579.

Thomson, M. C., Connor, S. J., Zebiak, S. E., Jancloes, M. and Mihretie, A. (2011). Africa needs climate data to fight disease. *Nature*, 471, 440–442.

Thomson, M. C., Mason, S. J., Phindela, T. and Connor, S. J. (2005). Use of rainfall and sea surface temperature monitoring for malaria early warning in Botswana. *American Journal of Tropical Medicine and Hygiene*, 73, 214–221.

UNISDR – United Nations International Strategy for Disaster Reduction (2003). *Regional Report on Early Warning of Natural Disasters in Africa*. Geneva, Switzerland: United Nations International Strategy for Disaster Reduction.

UNISDR – United Nations International Strategy for Disaster Reduction (2007). *Hyogo Framework for Action 2005–2015: Building the Resilience of Nations and Communities to Disasters*. Geneva, Switzerland: United Nations International Strategy for Disaster Reduction.

UNISDR – United Nations International Strategy for Disaster Reduction (2011). *Global Assessment Report on Disaster Risk Reduction*. Geneva, Switzerland: United Nations International Strategy for Disaster Reduction.

WHO – World Health Organization (2001). *Malaria Early Warning Systems, Concepts, Indicators, and Partners: A Framework for Field Research in Africa*. Geneva, Switzerland: World Health Organization.

Williams, A. P. and Funk, C. (2011). A westward extension of the warm pool leads to a westward extension of the Walker circulation, drying eastern Africa. *Climate Dynamics*, 37, 2417–2435.

Part I

Case Studies from North America

A Few Words on Andrew W. Garcia

Dr Andrew W. Garcia worked for the US Army Corps of Engineers in Vicksburg, Mississippi, for 39 years, and in 2008 joined the Corps Action for Change team to help revise policies and procedures for adapting to the effects of climate change.

Tragically, Andrew passed away suddenly while traveling in Canberra, Australia, shortly after completing his chapter on Hurricane Katrina. We are grateful to Andrew's colleague Dr Kathleen White for assisting us to finalise the chapter and to his family for allowing us to include Andrew's paper in the book. Kate gives a few words on Andrew and his career:

Andrew's curiosity and continual desire to learn led him on a personal and distinctive journey from physics to physical oceanography to atmospheric science. With this background, Andrew excelled as a scientist in an engineering organisation, as a meteorologist in a hydraulics laboratory, and finally as an interagency and international collaborator around sometimes-controversial climate change issues. This speaks not only for his technical and scientific competence, but also for his gentle perseverance and unwavering drive to understand issues in a way that could be used to better people's lives. Andrew put his multidisciplinary knowledge to good use in the days and months following Hurricane Katrina, to understand important issues around the tides, datums, waves, surges, and finally the hurricane itself. In recognition of his work on the post–Katrina Interagency Performance Evaluation Taskforce, Andrew was awarded the Department of the Army Superior Civilian Service Award and the US Army Corps of Engineers' Chief of Engineers Medal for Excellence in 2008. Without Andrew's unique combination of skills and his calm, thoughtful, reasonable, gently persuasive personality, the United States would be much less well-prepared for the future changes to come.

3

Hurricane Katrina and the City of New Orleans

ANDREW W. GARCIA*

Hurricane Katrina is perhaps the most well-documented landfalling tropical cyclone to date and one of the most catastrophic natural disasters in US history. From its genesis in the western Atlantic to its first landfall in southern Florida and second landfall in southern Louisiana, it was observed by satellite imagery, reconnaissance aircraft, moored ocean buoys and ships of opportunity (Knabb et al., 2006). Days prior to Katrina's landfall, news media were speculating that it could be the realisation of the New Orleans 'doomsday scenario' (Ringle, 2005). The doomsday scenario assumes a major (Saffir-Simpson Scale Category 4 or 5) cyclone on a track that would bring eye landfall just west of New Orleans and would push Gulf of Mexico waters into Lake Pontchartrain.

Most of the ground elevations of what is now the city of New Orleans proper, with the exception of the French Quarter, lie below the surfaces of Lake Pontchartrain and the Mississippi River, both of which are adjacent to the city. The south shore of Lake Pontchartrain lies on the north side of the city, where the city is protected only by a network of man-made levees. The concern was the possibility of Lake Pontchartrain waters, forced by Katrina's winds, overtopping or breaching the Pontchartrain levee system, thereby flooding New Orleans. In the end, New Orleans was flooded, but not because of failure of the Pontchartrain levee system.

3.1 Katrina Meteorology

Hurricane Katrina formed in the western Atlantic as a tropical depression near the Bahama Islands on 23 August 2005. By the morning of 24 August, it had acquired cyclone characteristics and was named Katrina at about noon UTC with an estimated centre position about 120 km from Nassau, Bahamas. Katrina intensified rapidly during the next twenty-four hours and attained hurricane status on 25 August and

* Deceased.

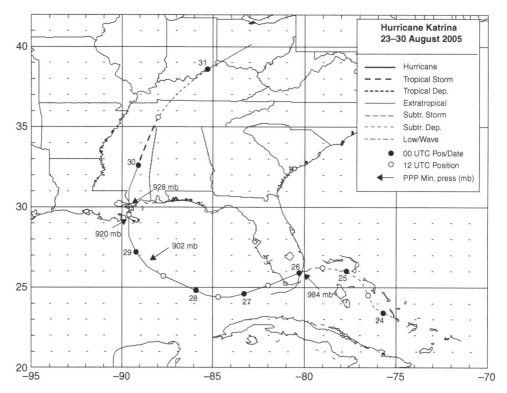

Figure 3.1 Hurricane Katrina best track (Knabb et al., 2006; National Oceanic and Atmospheric Administration).

made initial landfall along the south-eastern Florida coast late on 25 August as a Category 1 hurricane (maximum wind speed of ~35 m/s; 78 mph) on the Saffir-Simpson Scale (Figure 3.1). Katrina emerged from the south-western coast of Florida into the south-eastern Gulf of Mexico early on 26 August as a tropical storm with maximum wind speed of ~30 m/s (67 mph).

Once back over warm Gulf of Mexico waters, Katrina quickly regained hurricane strength. On 28 August, Katrina underwent a period of very rapid intensification from a minimal Category 3 hurricane to a Category 5 hurricane in only twelve hours with maximum wind speed of ~73 m/s (163 mph). About six hours later, Katrina attained its maximum intensity of ~75 m/s (168 mph) wind speed when located about 315 km south-east of the mouth of the Mississippi River. At this time, hurricane-force winds extended up to 165 km from the centre, making Katrina not only a very intense but also a very large hurricane. Katrina then took a more northerly track and, because of a series of eye-wall replacement cycles and possible dry air entrainment, weakened to a strong Category 3 hurricane before making landfall near Buras, Louisiana, then crossing the Mississippi River Delta to make final landfall near the mouth of the Pearl River on 29 August at approximately 09:45 local time with estimated maximum winds of 53 m/s (119 mph).

Katrina's weakening prior to landfall and final track slightly to the east of New Orleans avoided what has been described as the 'New Orleans Doomsday Scenario' in which a Category 5 hurricane would pass New Orleans just to the west, placing the city on the strong side of the cyclone (Grunwald, 2006). Because Katrina passed to the east as a Category 3, initial impressions were that New Orleans had escaped disaster.

There are no complete anemometer records in the immediate New Orleans vicinity, so maximum wind speeds can only be inferred. The National Hurricane Center (NHC) best track report (Knabb et al., 2006) suggests that New Orleans proper did not experience winds greater then Category 2. The closest complete anemometer record was obtained at a National Weather Service station located in Lake Pontchartrain, which recorded a maximum sustained wind of 35 m/s (78 mph). This anemometer location is about 13 km north of the south shore of Lake Pontchartrain and has over-water exposure and therefore tends to support the NHC best track wind estimate.

3.2 Katrina Surge Effects

Katrina generated the highest (8.5 m) and most extensive storm surge recorded in United States history, primarily along the Mississippi coastline (IPET, 2007a). Precise water-level observations are not available because all tide stations in the immediate vicinity of landfall were destroyed. Surge level estimates were based on high water marks recovered post-event and were likely influenced by wind wave effects. Figure 3.2 shows computed contours of the envelope of maximum surge levels as estimated using a numerical surge model (IPET, 2007a). Although Katrina made landfall as a Category 3 hurricane, the record high surge level is attributed to fact that Katrina was a very large Category 5 hurricane only a day before landfall. The National Data Buoy Center (NDBC) buoy station 42040 located about 118 km south of Dauphin Island, Alabama, recorded a maximum significant wave height of 16.91 m, the greatest significant wave height recorded by a Gulf of Mexico NDBC buoy. Although NDBC buoys do not record maximum individual wave heights, a statistical estimate of the maximum wave height was about 32 m (IPET, 2007a).

The New Orleans area did not experience the magnitude of surge heights as did the western Mississippi coast. However, to the east of New Orleans in St. Bernard and Plaquemines Parishes, storm surge elevations of 4.6 m to 5.8 m were estimated based on high-water marks (IPET, 2007a). It was along this section of the levees that Katrina-generated surge exceeded the design levee elevation criteria, which were between 3.6 m and 4.6 m (IPET, 2009). Figure 3.3 shows a detail of contours of the envelope of maximum surge levels as estimated using a numerical surge model (IPET, 2007a).

Property damage to the Mississippi coast residential units alone in the three coastal counties is estimated at approximately US$1.3 billion, as most structures within a

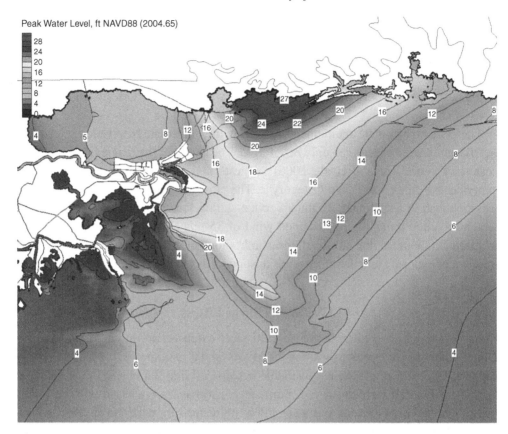

Figure 3.2 Computed maximum envelope of Katrina storm surge (courtesy of US Army Corps of Engineers).

couple hundred meters of the affected shoreline were completely destroyed. Total insured damages caused by Katrina for the United States are estimated at US$40.6 billion by the American Insurance Services Group, and the value of uninsured losses is estimated to be comparable to those insured (Knabb et al., 2006).

3.3 The Hurricane Protection System

A schematic overview of the New Orleans Hurricane Protection System (HPS) is shown in Figure 3.4 along with locations of the de-watering pumping system (IPET, 2009). The 'system' is, in reality, a combination of three hurricane protection projects, the first initiated with the Flood Control Act passed in 1965 by the United States Congress. Because of the piecemeal construction and incompleteness of the three projects, the system never attained the integrity for which it was designed. From the inception of the HPS project, delays and cost increases occurred as a result of technical issues, environmental concerns, legal challenges, and local opposition to various aspects of the project. Although the HPS was authorised by the US Congress

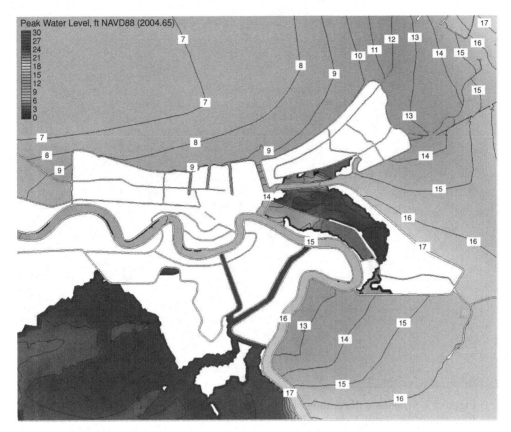

Figure 3.3 Detail of computed maximum envelope of Katrina storm surge (courtesy of US Army Corps of Engineers).

in 1965, a final plan to provide protection to the greater New Orleans area was not accepted until 1992. At no time did the entire New Orleans area and vicinity have a reasonably uniform level of flood protection. Indeed, some parts of the HPS were not scheduled for completion until 2015 (IPET, 2009).

The HPS design was based, in part, on the characteristics of a Standard Project Hurricane (SPH), which is approximately equivalent to a fast-moving Category 3 hurricane (the Saffir-Simpson Scale did not exist at the time). The SPH criteria were developed by the (then) US Weather Bureau and are loosely based upon the characteristics of hurricane Betsy, which badly damaged New Orleans in 1965. The characteristics of the SPH were intended to represent the most severe conditions likely to be experienced in the south Louisiana coastal area, with a maximum landfall cyclone wind speed of about 45 m/s and estimated 1 per cent annual recurrence probability. In 1979, the National Oceanic and Atmospheric Administration published a report that significantly revised the SPH criteria (Schwerdt et al., 1979). The revised criteria became the basis for the portion of the HPS that was designed subsequent to

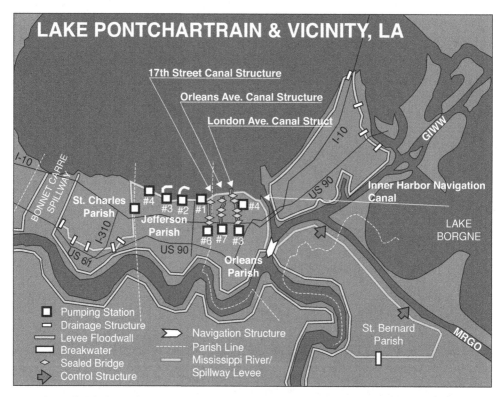

Figure 3.4 Schematic illustration of New Orleans pumping station and levee systems (courtesy of US Army Corps of Engineers).

1979. Other portions of the HPS continued to use the original SPH criteria until the time of Katrina.

Oddly, the de-watering pumping system depicted in Figure 3.4 was not a part of the HPS. Construction of the de-watering system long preceded the HPS, and it was designed to deal with local precipitation and seepage through the surrounding levee system but not with overtopping of the levee system. Fronting structures to protect the pumping stations from flooded outfall canals were an integral part of the HPS design. However, the fronting structures for pumping stations on the Orleans and London Avenue canals were never completed because of lack of funding. This left the Orleans Avenue Canal with a section of legacy floodwall with a lower elevation than adjacent floodwalls and a direct route for water to enter the city without overtopping the adjacent floodwalls.

3.4 The Consequences

At the time of Katrina, the New Orleans population was approximately 455,000. More than 80 per cent of the metropolitan New Orleans area was flooded, in places to depths exceeding 4.5 m. The official death toll in the five parishes comprising New Orleans was 1,464 (Louisiana Department of Health and Hospitals, 2006). More than 75 per cent of the fatalities were those over 65 years of age. Direct property losses are estimated to exceed US$20 billion plus another US$7 billion in public structures and utilities. Post-event analysis concluded that even if the HPS had functioned under the best possible scenario, there would have been US$10 billion in damages not including the damage to public utilities (IPET, 2009). Estimation of damages from Katrina was complicated by the passage of major hurricane Rita, which occurred only three weeks after Katrina. Rita made landfall to the west of New Orleans, but precipitation from Rita re-flooded some areas and further weakened the already damaged levees.

Because of the prolonged duration of lack of public infrastructure (e.g. water, electrical power, food supplies, medical facilities), many evacuees could not or chose not to return to the area. As an example of the collapse of services, in St. Bernard Parish, the number of pre–Katrina electric utility customers was 29,145. Post–Katrina (December 2005), the number was 178. Follow-on studies of re-population of New Orleans through 2006 showed a close correlation of the depth of flooding to per cent of re-population, with areas of flood depths less than 0.6 m experiencing re-population rates of about 75 per cent and areas of flood depths greater than 1.2 m experiencing re-population rates of about 10 per cent (IPET, 2007b). Many families that evacuated were forced to re-locate significant distances from New Orleans but then gained employment and established residences elsewhere and had no motivation to return. Next to the tragic loss of life, the breakdown in social structure was perhaps the most significant consequence of Katrina.

Beyond the immediate effects to built and social infrastructure, Katrina also caused significant environmental damage. The most serious was saltwater flooding of wetlands and marshes along with extensive freshwater fish and oyster kills. The United States Geological Survey estimated that more than 295 km^2 of wetland habitat was lost from the combined effects of hurricanes Katrina and Rita. This is four times the average annual loss for the area of about 70 km^2 measured since the 1930s.

3.5 Findings and Lessons Learned

Following the flooding of New Orleans caused by Katrina, the US Army Corps of Engineers, the agency responsible for design and construction of the system, assessed the hurricane protection system in New Orleans and southeast Louisiana as '. . . a system in name only' (IPET, 2009). Although the HPS generally was built as designed, the system did not perform as a system. Some areas of the system were incomplete, and much of the system was below specified elevations. Moreover, because it is based upon the characteristics of a single event, the SPH methodology used to develop the design criteria did not depict the true vulnerability of the area. This was not apparent at the time because the overall process used to develop risk and flooding vulnerability estimates produced reasonable results when compared to losses from historical events.

Local land subsidence contributed to the surge effects being significantly greater than the design assumptions, particularly for portions of the floodwalls. In some cases, the actual elevations were more than 0.6 m lower than design specifications. In addition to the overall subsidence problem, the elevation of one portion of the 17th Street Canal was based upon an inadequately documented bench mark monument. The top of the levee floodwall was at the correct design elevation relative to the bench mark, but the bench mark's actual elevation was almost 0.6 m below its reported elevation. This error emphasised that accurate topographic control in project design and construction is essential to determine correct elevations.

3.6 Societal Effects

There were social factors in addition to the project design and construction deficiencies that contributed to the human toll. The risk level of potential inundation because of cyclones affecting the New Orleans area was neither well communicated to, nor understood by, the public. While many residents of the Mississippi Gulf coast experienced total destruction of their homes, they were able to evacuate and relocate elsewhere in a timely manner. Many New Orleans residents, on the other hand, were subjected to prolonged destruction and disruption from flooding caused by broken levees, subsequent flooding caused by hurricane Rita, and efforts to drain the city (Weems et al., 2007).

Katrina is commonly thought to have caused a reverse in the New Orleans population trend, but in fact the population of the city has been in decline since 1960. Katrina likely increased the negative population trend, and the 2010 population was estimated at about 343,000, about the same as in 1910. Moreover, the trend appears to be continuing independent of whether neighbourhoods were flooded by Katrina or not (Plyer, 2011a).

In contrast to the continued population trend pre– and post–Katrina, recent census data suggest that demographics in post–Katrina New Orleans are undergoing a significant shift. Prior to 2000, the average Hispanic population growth rate in the United States exceeded that of New Orleans by more than a factor of five. Between 2000 and 2010, the New Orleans Hispanic population growth has exceeded the United States average by almost 15 per cent (Plyer, 2011b). Moreover, unlike the United States at large, the influx of Hispanics into New Orleans is mostly composed of non-Mexican nationality, principally from Latin America. Anecdotal accounts suggest that many of the Hispanic immigrants are employed in service-sector jobs that were abandoned by local workers after Katrina.

Immediately after Katrina, local officials proclaimed that New Orleans would be re-built ' . . . bigger and better . . . ' Given the declining population, it is difficult to conclude that a future New Orleans will be bigger. Although 'better' is a subjective term, the number of New Orleans households with children under 18 years of age is declining faster than the national average (Plyer, 2011b). This implies a local school system with declining enrolment, hardly a trend that fosters a healthy, better-balanced community.

A study conducted by the National Academy of Sciences on the reconstruction of New Orleans observed that disasters tend to accelerate existing economic, political, and social trends, and immediately following Katrina there was a rush by residents to re-build the familiar rather than consider ways to betterment. Moreover, parallel and conflicting planning processes at the state and city levels during the first year post–Katrina led to contrasting planning goals. The study concluded that post–Katrina New Orleans is no exception to similar disaster trends, that the pre–Katrina trends will persist, and the prospects for more than partial recovery are bleak (Kates et al., 2006).

3.7 Concluding Remarks

While it is tempting to view hurricane Katrina from an anthropomorphic perspective as a malevolent event, Katrina was in reality simply the agent of disaster. That the New Orleans area would at some time be struck by a major hurricane had been predicted for years; the only question was when. The human deaths and injuries and the destruction and damage to property were the results of decades of unstructured land development, political disputes and mismanaged planning of the protection system, piecemeal funding and construction of the protection system and obsolete methodology for estimating the cyclone risk. Despite upgrading and enhancement of the HPS, population control and evacuation are probably the only credibly effective means to substantially reduce the risk of loss of life in a similar future event (Link, 2010).

References

Grunwald, M. (2006). The Army Corps of Engineers is the real culprit behind New Orleans' devastation. *Grist Magazine*, Seattle, 29 August 2006. Accessed 15 October 2012 from: <http://www.grist.org/article/grunwald>.

IPET – Interagency Performance Evaluation Task-force (2007a). *Performance Evaluation of the New Orleans and Southeast Louisiana Hurricane Protection System. Volume IV, The Storm*. Washington, DC: US Army Corps of Engineers.

IPET – Interagency Performance Evaluation Task-force (2007b). *Performance Evaluation of the New Orleans and Southeast Louisiana Hurricane Protection System. Volume VII, The Consequences*. Washington, DC: US Army Corps of Engineers.

IPET – Interagency Performance Evaluation Task-force (2009). *Performance Evaluation of the New Orleans and Southeast Louisiana Hurricane Protection System. Volume I, Executive Summary and Overview*. Washington, DC: US Army Corps of Engineers.

Kates, R. W., Colten, C. E., Laska, S. and Leatherman, S. P. (2006). Reconstruction of New Orleans after Hurricane Katrina: a research perspective. *Proceedings of the National Academy of Sciences of the United States of America*, 103, 14653–14660.

Knabb, R. D., Rhome, J. R. and Brown, D. P. (2006). *Tropical Cyclone Report, Hurricane Katrina, 23–30 August 2005*. Florida, USA: National Hurricane Center.

Link, L. E. (2010). The anatomy of a disaster, an overview of Hurricane Katrina and New Orleans. *Ocean Engineering*, 37, 4–12.

Louisiana Department of Health and Hospitals (2006). *Reuniting the Families of Katrina and Rita: Final Report of the Louisiana Family Assistance Center*. Baton Rouge: Louisiana Department of Health and Hospitals.

Plyer, A. (2011a). *Population Loss and Vacant Housing in New Orleans Neighbourhoods*. New Orleans, LA: Greater New Orleans Community Data Center.

Plyer, A. (2011b). *Homeownership, Household Makeup, and Latino and Vietnamese Population Growth in the New Orleans Metro*. New Orleans, LA: Greater New Orleans Community Data Center.

Ringle, K. (2005). In Camille's deadly 1969 solo, a grim prologue to Katrina. *The Washington Post*, Washington, DC, 29 August 2005. Accessed 15 October 2012 from: <http://www.washingtonpost.com/wp-dyn/content/article/2005/08/28/AR2005082801050.html>.

Schwerdt, R. W., Ho, F. P. and Watkins, R. R. (1979). *Meteorological Criteria for Standard Project Hurricane and Probable Maximum Hurricane Windfields, Gulf and East Coasts of the United States, NOAA Technical Report NWS 23*. Silver Spring, MD: National Weather Service, National Oceanic and Atmospheric Administration, US Department of Commerce.

Weems, C. F., Watts, S. E., Marsee, M. A. et al. (2007). The psychosocial impact of Hurricane Katrina: contextual differences in psychological symptoms, social support, and discrimination. *Behaviour Research and Therapy*, 45, 2295–2306.

4

A Brief History of Flooding and Flood Control Measures Along the Mississippi River Basin

TIMOTHY M. KUSKY

Stream and river valleys have been preferred sites for human habitation for millions of years, dating back to our earliest known ancestors in Turkana Gorge in the Great Rift Valley of Kenya (e.g. Kusky, 2008). These locations offer routes of relatively easy access through rugged mountainous terrain, provide life-sustaining drinking water for people and livestock and are invaluable for irrigation. The soils in river valleys are also some of the most fertile that can be found because they are replenished by yearly or less frequent floods. The initial wave of European settlers to America flocked to valleys carved by the Hudson, James (Virginia) and Ohio rivers. As the country grew, Americans trekked west to settle along the Mississippi, Missouri, Red and many other waterways. Urban centres sprang up along their courses, for rivers provided convenient, quicker and cheaper transportation than overland travel. But human habitation and progress have left a dark legacy. Many, if not most, of the streams and rivers are polluted because industry has discharged billions of gallons of chemical waste into the waterways. Humans themselves have shown an almost callous disregard for rivers by allowing harmful agricultural runoff laced with pesticides and herbicides or simply dumping trash and household waste into the water.

Stream and rivers are dynamic environments, and their benefits do not come without a price. Their banks are prone to erosion, and the rivers periodically rise above their borders. During floods, rivers typically cover their floodplains with 5 to 10 feet (1–3 m) of water and deposit layers of silt and mud. This is part of a river's natural cycle and was relied upon by generations of farmers for replenishing and fertilising their fields. Human needs precipitated a dramatic change in the past century regarding floodplain use, which upset the natural state. Thousands of miles of earthen levees, flood walls and river control structures have been built along many of the nation's and world's rivers, while local legislation allowing extensive development of these floodplains has crept into place in many states (IFMRC, 1994). An example of the effects of blocking off the floodplains from the river and increased urbanisation of the floodplain is provided by the Lower Missouri River.

The Missouri River winds its way over 2,300 miles (3,700 km) and drains about one-sixth of the United States (Hesse and Sheets, 1993). It was once one of the wildest stretches of river in the American Midwest. During the past two centuries, the Missouri River, along with its adjacent wetlands and floodplains, has been dramatically modi-fied in various attempts to promote transportation, agriculture and development. These modifications have included draining wetlands for cultivation, straightening stream channels to facilitate navigation, stabilising banks to prevent erosion and construct-ing agricultural levees, dams, reservoirs and flood control structures to regulate flow and exclude excessive waters from the floodplain (IFMRC, 1994; Pinter and Heine, 2005). These changes have produced a negative impact resulting in a severe loss of vital wetlands. Historically, the Missouri River floodplain below Sioux City, Iowa, covered 1.9 million acres. In a court case, Sierra Club USA argued that alterations to the river floodplain system have resulted in the loss of approximately 168,000 acres of natural channel, 354,000 acres of meander belt habitat and 50 per cent of the Mis-souri River's surface. In addition, shallow-water habitat has been reduced by up to 90 per cent in some areas while sandbars, islands, oxbows and backwaters have been virtually eliminated. Forested floodplains along the Missouri River have decreased from 76 per cent in the nineteenth century to 13 per cent in 1972, while cultivated lands increased from 18 per cent to 83 per cent during the same time (Case 03–04254-CV-C-SOW of *Sierra Club vs. US Army Corps of Engineers*). By the late 1970s, the Lower Missouri River had been totally channelised and its natural floodplain eco-systems almost completely converted to agricultural or other uses. Today, the Lower Missouri River is flanked by levees and other flood control measures for most of its length.

As the floodplains are cordoned off by levees and towns are built in the floodplain, the ability of the floodplain to perform its natural functions decreases. Soils are no longer replenished and nourished by annual flooding. Floodplains can no longer absorb water from spring floods, thus reducing the ability to control the severity of floods. Many critical natural ecosystems are threatened, and some species of fish along with other fauna and flora that rely on the annual flood cycle on the floodplains have become endangered and face the alarming possibility of extinction. People residing within the floodplain face a precarious future as well. Many residents of floodplains are living with a false sense of security, because they are told and believe that the levees are safe and that their homes and businesses are not threatened by high water. Stark examples highlight this misguided hope. Extended floods such as in 1973, 1993, and 2008 along the Mississippi River document that this security is not certain. During the 1993 floods, nearly 80 per cent of the private levees along the river were overtopped, breached or outright failed, inflicting billions of dollars of damage (Criss and Wilson, 2003). Since then, many areas on the floodplains, such as the Chesterfield Valley in Missouri along the Missouri River, have experienced increasing development on the floodplain in areas that were under 5 to 10 feet (1–3 m) of water in the 1993 floods

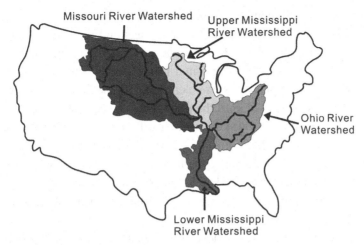

Figure 4.1 Map of the Mississippi and Missouri river systems (modified from US Geological Survey, 2012).

and have experienced more development since this flood than in the entire previous history of the region (Kusky, 2008). Global climate change models suggest that the central plains of the United States may enter a much wetter climate phase within the next 30 years (Wuebbles and Hayhoe, 2004; Pan et al., 2009), dramatically increasing the frequency and intensity of floods (e.g. Criss, 2009).

Today, because many floodplains are industrialised or populated by residential neighbourhoods, the floods are no longer welcomed as bearers of renewal for the environment; rather, they are regarded as disasters instead of beneficial fertilisation events. On average, floods kill scores of people annually in the United States and cause billions of dollars in crop and property damage (e.g. Kusky, 2008). Dikes and levees have been built around many rivers in attempts to prevent floodwaters from invading urban areas. But the perceived cure can be worse than the problem. These types of structures confine the river to a narrow channel, and the waters rise more quickly and cannot seep into the ground of the floodplain (Belt, 1975; Pinter et al., 2000; Pinter and Heine, 2005). Flood stages are systematically higher for the same amount of flow in the rivers following river constriction and floodplain urbanisation (Criss and Shock, 2001).

4.1 History of Levee Building and Floods on the Mississippi River

The Mississippi River is the longest river in the United States, the third longest in the world (Figure 4.1), and encompasses the third largest watershed, draining 41 per cent of the continental United States, including an area of 1,245,000 square miles (3,224,550 km^2). It has a wide range of physiographic characteristics and biological habitats along its length (US Geological Survey, 2012). The river transports

230 million tons of sediment, including the sixth largest silt load in the world (e.g. Kusky, 2008). Before the arrival of Europeans and their altering of the river, this silt would cover and fertilise the floodplains during the semi-annual floods and carry even more downriver to be deposited in the river's delta. Levee construction along the Lower Mississippi River system ensued with the first settlers who migrated to the region and has continued until the present-day levee system, the main parts of which include 2,203 miles (3,580 km) of levees, floodwalls and other control structures. Of this, 1,607 miles (2,586 km) of levees lie along the Mississippi River, and another 596 miles (959 km) are along the banks of the Arkansas River in Arkansas, Oklahoma, Kansas and Colorado and along the Red and Atchafalaya rivers in Louisiana. Additional levees are in place along the Missouri, Ohio and Illinois rivers (Hesse and Sheets, 1993).

The first levee to appear along the Mississippi River was built around the first iteration of New Orleans between 1718 and 1727 and consisted of a slightly more than a mile-long (5,400 feet; 1,646 m), 4-feet (1.2 m) -high earthen mound that was 18 feet (5.5 m) wide at the top, with a road along the crown. The design was meant to protect residents and property of the newly founded city from annual floods and accompanying pestilence that could last for several months from March until June of each year. Founded in 1718, even in its infancy, New Orleans offered great promise. Following Sieur de La Salle's 1682 exploration of the Mississippi River, King Louis XIV's Secretary of the Navy, Louis, Comte de Pontchartrain, commissioned the explorer Pierre Le Moyne, Sieur d'Iberville to secure the region for France, possibly to establish a fort or colony near the mouth of the Mississippi River but most importantly to prevent any Spanish incursions. It would be d'Iberville's younger brother, Jean Baptiste Le Moyne, Sieur de Bienville, who founded La Novelle-Orleans. He chose the site carefully. First, it was an ideal location to control the portage between the Mississippi River and what is today known as Lake Pontchartrain. Also, the new settlement could be instrumental in France's control of the lumber and fur trade moving downriver. Finally, its very location indicates the founders were keenly aware of the potential for severe flooding. Then and now, New Orleans resembles an inland island surrounded by water on all sides. Lakes Pontchartrain, Maurepas, Borgne, Bayou Manchac and the Amite River divide it from higher land on the north, and the Mississippi River wraps around its other sides. The site of New Orleans on the natural levee of the river on the Isle of Orleans has always been tenuous, and the city has been inundated by floods from the river on three sides and by storm surges from hurricanes on the other side about every 30 years since its founding (Kusky, 2008). The first levee built in 1718 to 1727 proved inadequate to protect the city from rampant flooding. On 23 and 24 September, 1722, a devastating hurricane roared in from the Gulf of Mexico, nearly completely destroying the newly designated capital city of French Louisiana. The storm battered the city with 100-mile-per-hour (161 km/hr) sustained winds coupled with a storm surge of 7 to 8 feet (2–2.4 m) that overtopped the 4-feet

(1.2 m) -high levee. Almost every structure within the supposed safety of the levee was destroyed or severely damaged. If city planners had taken this warning when the New Orleans consisted of just a few dozen buildings, much damage could have been avoided in the future. Instead, more and higher levees were raised and the result was the same as successive storms unleashed floods destroying or severely damaging the city in 1812, 1819, 1837, 1856, 1893, 1909, 1915, 1947, 1956, 1965, 1969 and 2005. The old levees did not hold in 1722, and the new and improved levees shared the same fate in 2005 from the onslaught of hurricanes Katrina and Rita. But not just levees were employed to combat flooding. Canals both above and below ground were proposed to move water out of the city rapidly, and in the 1830s a novel idea was suggested – instillation of large water pumps. Though this approach is more suited to remove water from the city accumulated during torrential downpours than to offer flood protection, it was thought to certainly be of some use nonetheless. The first editions of these pumps were steam powered, later electrified with diesel generators as backup power sources. By the 1980s, a remarkable array consisting of 20 pumping stations armed with 89 pumps laced through New Orleans and provided a sense of security. The residents' feelings were seemingly justified. These massive machines could remove an astonishing 15,642,000 gallons (59,210,000 litres) of water per minute, equivalent to the flow of the Ohio River. As time would prove soon enough, the Ohio River pales compared to the force of a storm the magnitude of hurricane Katrina. But levees remained the flood defence of choice not only for New Orleans but also for the entire length of the Mississippi River.

Early river levees along the Lower Mississippi consisted of earthen mounds, generally with a slope of 1:2. The local and state governments declared a policy that resident land holders had to build their own levees along the property they owned edging the river. Haul methods for bringing the dirt to make the levees were primitive by today's standards, typically with horse and carriage, yielding a paltry 10 to 12 cubic yards (7.5–9 m^3) per day. The federal government became involved in 1820 with legislation that focused mostly on river navigation and did not consider flood control as a priority. As levees were built at a break-neck pace, the river became constricted, and this caused sections of the riverbed to continuously raise themselves in a process called aggradation. This occurs when the river is not allowed to migrate laterally; it cannot move out of the way of the sediment it is carrying and depositing and is unable to widen the channel, therefore raising the bed as this sediment is deposited. The final control on the height of the riverbed is, however, the downstream base level of the ocean. Disastrous floods along the Lower Mississippi River in 1844, 1849 and 1850 resulted in passage of the Swamp Acts of 1849 and 1850. This legislation empowered the states of Louisiana, Mississippi, Arkansas, Missouri and Illinois to identify floodplains and wetlands that were unfit for cultivation. These lands were sold, and the revenues generated were used to construct levees and to drain the purchased lands for cultivation. Between 1850 and 1927, levees bordering the Lower Mississippi had to be continuously heightened because of the river avulsion, a consequence from the construction of the levees.

In 1850, Congress appropriated US$50,000 to complete two topographic and hydrographic surveys in an effort to promote flood protection along the Mississippi River. One survey was completed by a civilian engineer, Charles Ellet Jr, and the other by Army Engineers A. A. Humphreys and Henry Abbot. The Humphreys-Abbot Report recommended three possible methods for flood control, including cutting off the bends in the river, diversion of tributaries creating artificial reservoirs and outlets and confining the river to its current channel using levees. Since the first two options were considered too expensive, the third was selected and would prove to be the harbinger of long-lasting negative consequences. Their levee design called for freeboards at 3 to 11 feet (1–3 m) above the high water mark of the 1858 flood.

Not surprisingly, during the American Civil War, from 1861 to 1865, the levees fell into a state of disrepair, and this was only made worse by the large floods of 1862, 1865 and 1867. The floods of 1874 prompted the creation of the Levee Commission to complete a new levee survey and make recommendations for how to repair the system and reclaim the floodplain. The Levee Commission made a stark assessment, citing major defects in the system and enormous costs for repair and improvement. They documented that previous levees were built in unsuitable locations, dogged by poor organisation, insufficient height, sub-standard construction and inadequate inspection and guarding. They estimated that it would cost US$3.5 million to repair the existing system and US$46 million to build a new, complete levee system to reclaim the floodplain. These were staggering sums in 1870s American dollars for the government to allocate.

In 1879, the Mississippi River Commission (MRC) was created by Congress, as organised by James B. Eads. The Commission consisted of three officers from the US Army Corps of Engineers, three civilians and one officer from the US Coast and Geodetic Survey. The MRC conducted surveys and suggested numerous modifications and new additions to the flood control and navigation projects along the river. During 1882, the levees were improved and raised to contain the frequent floods, but they would not hold for long.

The flood of 1890 destroyed 56 miles (90 km) of levees, and the MRC began to raise them even higher from 38 to 46 feet (11.5–14 m). During this phase of massive reconstruction, the federal government as well as private citizens added more than 125 million cubic yards (96 million m^3) of soil to the levees, but much of this work was lost to the river by mass wasting processes including slumping and bank caving. Efforts were made to reinforce the banks with various revetments, but then the flood of 1912 struck, once again laying waste to much of the levee system that was intended to protect the adjacent floodplain. The commission proposed its typical response – raise the levees again, this time by a meagre 3 feet above the 1912 flood line. Continued aggradation of the riverbed with each increase in levee heights was at the time not recognised as a growing threat.

The inaugural Federal Flood Control Act was passed in 1917, authorising for the first time levees to be built exclusively for flood controlling the Mississippi main-stem and its tributaries. The federal government would pay two-thirds of the levee costs if the local interests contributed the remainder. During the 1920s, levee construction was stepped up to a higher pace with the mechanisation of earth-moving technology and the introduction of large cranes, moving tower machines and cableway draglines that could move dirt orders of magnitude faster than the out-of-date horse and cart.

The year 1927 came, and with it the greatest flood in recorded history along the lower Mississippi River Valley (Belt, 1975). The stage was set with heavier-than-normal rains in the summer of 1926, followed by intense winter storms and heavy rains in early 1927 that continued through the spring and early summer. Many of the levees built to the MRC standards failed up and down the river, with enormous consequences in terms of loss of life, displaced people and loss of property that was supposed to be protected by the levees. Some areas along the route of destruction saw the river escape its natural and man-made confines and spread 50 miles (80 km) wide. Below Memphis, Tennessee, the width of the river was around 60 miles (97 km). The entire area ravished by the flood exceeded 27,000 square miles (70,000 km^2). The government quickly responded with the 1928 Flood Control Act, enacting legislation to improve the grade of the levees and develop models of various flood scenarios, including the creation of several large floodways that could be opened to let water out of the river during high flow times. Some of these floodways were quite large, such as the Birds Point-New Madrid in Missouri, which is about 35 miles (56 km) long, 3 to 10 miles (5–16 km) wide and designed to divert 550,000 cubic feet (15,576 m^3) per second of flow from the Mississippi during floods. Further downriver, the West Atchafalaya Floodway was envisioned to carry half of the modelled projected flood of 1,500,000 cubic feet (139,400 m^3) per second. Upriver from New Orleans is the Bonnet Carré Floodway. Its purpose is to restrict the flow to downstream by diverting the water and protecting homes and businesses of more than 1.1 million people (in 2010) of the Greater New Orleans area. Levees were redesigned, moved to locations where their projected lifespan would be from 20 to 30 years and, of paramount importance, thought to be stronger. As the construction on the new levee and floodway system continued, additional floods, such as the 1929 flood, disrupted operations, but the construction methods continued to improve and the levees rose, forming much of the present-day system.

In January and February of 1937, heavy winter storms produced a large flood that emanated from the Ohio River watershed, raising the water levels to a point requiring the Birds Point-New Madrid Floodway to be used, opening the floodway by dynamiting the Fuse Plus Levee. Damage stretched from Pittsburgh, Pennsylvania, to Cairo, Illinois, killing 385 people and leaving another one million homeless. Property damage was estimated at US$500 million. Though at a terrible cost to lives and infrastructure, breaking of the Birds Point Levee released huge volumes of water and eased the flood downstream. One of the lessons gleaned from the 1937 flood was that roads should be added to the levees to aid in transporting material from place to place during floods. The MRC in 1947 began redesigning levees to be stronger to avoid future failure, recognising the importance of compaction for reducing the potential of calamity.

Levees fail when there is a large hydrostatic pressure difference across the surface caused by a significant difference in water height on either side. They fail by three main modes: Underseepage of water beneath the levee, where the pressure from the high water opens a channel causing catastrophic failure; hydraulic piping in which the water finds a weak passage through the levee; and overtopping, when the water flows over the top of the levee and erodes the sides (e.g. Kusky, 2008). They can also fail when the river current scours the base of the levee during high-flow conditions, as occurred time and time again during Mississippi River floods, causing slumping and massive collapse of the levee. Mass wasting is also promoted by long-term floods in which the water gradually saturates the pores of the levee, weakening it, developing massive liquefaction and catastrophic failure, leading large sections of the levee to collapse at the same time. Most levee failures happen during times when the flow has been high for long periods, because this increases the pore pressure, scouring and liquefaction potential of the levee.

By 1956, the MRC was modelling floods with twice the discharge as previous, examining the ability of the river and levee system to withstand a discharge of 3,000,000 cubic feet (2,300,000 m^3) per second. When the 1973 flood hit the Mississippi River basin, it was one of the highest floods recorded in 200 years (Belt, 1975; Noble, 1980; Pinter et al., 2000). It established a record for the number of days the river was out of bank, causing more than US$183.7 million in damages. In terms of flood management, the flood of 1973 brought about the realisation that building levees and wing dikes, as well as other navigational and so-called flood control measures, had actually decreased the carrying capacity of the river, meaning that for any given amount of water, the flood levels would be higher than before the levees were built (Pinter et al., 2000).

Enormous flooding in 1993 provided another test of the levees, and the new system failed massively. Constriction of the river caused by the levees led to numerous cases of structural failure, overtopping, crevasse splays, collapse and immense amounts of damage such as had never before been seen along the river. Approximately two-thirds of all the levees in the Upper Mississippi River basin collapsed or were breached or otherwise damaged by the floods of 1993. Dozens of people died and 50,000 homes were damaged or destroyed, with the total property loss estimate exceeding one billion dollars.

The Upper Mississippi River basin experienced record or near-record floods in June of 2008 (Holmes et al., 2010), rupturing at least 25 levees in Missouri, Iowa and Illinois (Figures 4.2, 4.3). Record-breaking floods in Iowa peaked at more than

Figure 4.2 Overtopped levee at Winfield, Missouri, June 2008 (photo by T. M. Kusky).

three metres above the previous mark at Cedar City, and the crest established new high water marks as it moved downstream to Burlington, Iowa. Numerous failures and overtopping of levees near Winfield, Missouri, worked to spread the water across the floodplain, lowering the crest to below record levels, before the flood reached the confluence with the Missouri River. Since the Missouri was not at record stages in June, the flood crest in the Middle Mississippi Basin was much lower than that of the Upper Mississippi.

Floods along the Mississippi and Missouri rivers returned in force in 2011, with the worst flooding on record since 1927 in many parts of the basin and new high levels achieved in many other locations. These floods illustrate that the vulnerability and present-day situation, even with all the flood control measures along the Mississippi, Missouri, Illinois and Ohio rivers, have not significantly reduced the threat of major flooding along the system. The floods began when, in April, two major storm systems dropped record amounts of rain throughout the Mississippi watershed on top of the springtime snowmelt in the channels. The US Army Corps of Engineers was forced to breach some levees, such as the Birds Point Levee in southern Missouri, flooding thousands of acres of farmland in the Birds Point-New Madrid Floodway but successfully lowering river levels and thus sparing cities such as Cairo, Illinois, from severe flooding. Two other floodways were opened, including the Bonnet Carré and Morganza spillways in Louisiana. This was the first flood in the history of the Mississippi during which the US Army Corps of Engineers was compelled to

Figure 4.3 A working floodplain, near Winfield, Missouri, 19 June 2008. The Cuivre River meanders through the centre of the photo, as outlined by submerged trees (photo by T. M. Kusky).

simultaneously open three flood control spillways to remove water from the river, overwhelming areas in the floodplains with up to 20 to 30 feet (6–10 m) of water. This example indicates how the levee system can protect floodplains from flooding for low-magnitude events, but in the ever-more-frequent high-magnitude floods with changing climate, the levee system cannot manage to contain the water and must be breached, either by failure or in controlled releases, to move the flood waters back to designated areas within the floodplain to protect other areas deemed more valuable by the US Army Corps of Engineers and other government authorities. Such decisions have proven to be a bitter and divisive issue among those communities left at the mercy of a raging, unforgiving river.

Acknowledgments

I thank Danny E. Butcher, Nicolas Pinter, Bob Criss and J. David Rogers for discussions on the history of flooding and flood control along the Mississippi and tributaries. Funds were provided by the C. B. Belt Foundation, National Natural Science Foundation of China (Grants 91014002 and 40821061) and the Ministry of Education of China (B07039 and 2011CB710600).

References

Belt, C. B. Jr (1975). The 1973 flood and man's constriction of the Mississippi River. *Science*, 189, 681–684.

Criss, R. E. (2009). Increased flooding of large and small watersheds of the central USA and the consequences for flood frequency predictions. In *Finding the Balance Between Floods, Flood Protection, and River Navigation*, ed. R. E. Criss and T. M. Kusky. St. Louis, MO: Centre for Environmental Sciences, St. Louis University, pp. 16–21. Accessed 15 October 2012 from: <http://ces.slu.edu/annualreport/FloodForum_Book_final.pdf>.

Criss, R. E. and Shock, E. L. (2001). Flood enhancement through flood control. *Geology*, 29, 875–878.

Criss, R. E. and Wilson, D. A. (eds.) (2003). *At the Confluence: Rivers, Floods, and Water Quality in the St. Louis Region*. St. Louis: Missouri Botanical Gardens Press.

Hesse, L. W. and Sheets, W. (1993). The Missouri River hydrosystem. *Fisheries*, 18, 5–14.

Holmes, R. R. Jr, Koenig, T. A. and Karstensen, K. A. (2010). *Flooding in the United States Midwest, 2008: US Geological Survey Professional Paper 1775*. Washington, DC: US Geological Survey. Accessed 15 October 2012 from: <http://pubs.usgs.gov/pp/1775/>.

IFMRC – Interagency Floodplain Management Review Committee (1994). *A Blueprint for Change. Sharing the Challenge: Floodplain Management into the 21st Century*. Washington, DC: Administration Floodplain Management Task Force.

Kusky, T. M. (2008). *Floods: Hazards of Surface and Groundwater Systems*. New York: Facts on File.

Noble, C. C. (1980). The Mississippi River flood of 1973. In *Geomorphology and Engineering*, ed. D. R. Coates. London: Allen and Unwin, pp. 79–98.

Pan, Z., Segal, M., Li, X.-Z. and Zib, B. (2009). Global climate change impact on the Midwestern USA – a summer cooling trend. In *Understanding Climate Change: Climate Variability, Predictability, and Change in the Midwestern United States*, ed. S. Pryor. Bloomington: Indiana University Press.

Pinter, N. and Heine, R. A. (2005). Hydrodynamic and morphodynamic response to river engineering documented by fixed-discharge analysis. Lower Missouri River, USA. *Journal of Hydrology*, 302, 70–91.

Pinter, N., Thomas, R. and Wlosinski, J. H. (2000). Regional impacts of levee construction and channelization, middle Mississippi River USA. In *Flood Issues in Contemporary Water Management*, eds. J. Marsalek, W. E. Watt, E. Zeman and F. Sicker. Boston: Kluwer Academic Publishers.

US Geological Survey (2012). *Upper Midwest Environmental Science Center*. La Crosse, WI: Upper Midwest Environmental Science Center. Accessed 16 October 2012 from: <http://www.umesc.usgs.gov/>.

Wuebbles, D. J. and Hayhoe, K. (2004). Climate change projections for the United States Midwest. *Mitigation and Adaptation Strategies for Global Change*, 9, 335–363.

5

The 2003 and 2007 Wildfires in Southern California

JON E. KEELEY, ALEXANDRA D. SYPHARD, AND C. J. FOTHERINGHAM

Although many residents of southern California have long recognised that wildfires in the region are an ongoing, constant risk to lives and property, the enormity of the regional fire hazard caught the world's attention during the southern California firestorms of 2003 (Figure 5.1). Beginning on 21 October, a series of fourteen wildfires broke out across the five-county region under severe Santa Ana winds, and within two weeks, more than 300,000 ha had burned (Keeley et al., 2004). The event was one of the costliest in the state's history, with more than 3,600 homes damaged or destroyed and twenty-four fatalities. Suppression costs for the 12,000 firefighters have been estimated at US$120 million, and the total response and damage cost has been estimated at more than US$3 billion (COES, 2004).

Just four years later, almost to the day, this event was repeated. Beginning on 22 October 2007, thirteen wildfires broke out across the same region, and under similar Santa Ana winds, consuming more than 175,000 ha, destroying more than 3,300 structures and killing seven people (Keeley et al., 2009). The 2003 and the 2007 wildfires were remarkably similar in their causes, impacts and the human responses they elicited. Particularly alarming is the observation that these fire events are not new to the region, as large fire events have occurred historically.

5.1 Prior Condition

Essentially every year, in all counties in the southern California region, there are fires that range in size from 1,000 to 10,000 ha (Keeley et al., 1999). This regional history of wildfires is largely a result of the Mediterranean-type climate, with winter rain growing conditions sufficient to produce dense vegetation and a long summer drought that converts this biomass into highly flammable fuels. Although these conditions occur periodically under other climatic regimes, the Mediterranean-type climate results in such conditions annually. Massive fires more than 50,000 ha, similar to the 2003 and 2007 fires, have occurred nine times since the earliest date for which we have records,

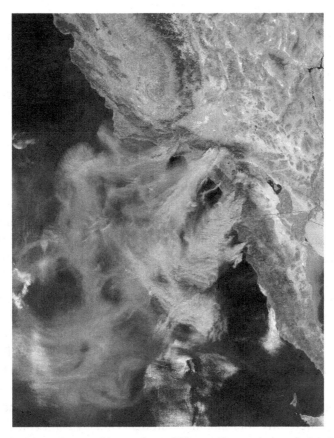

Figure 5.1 Smoke plumes blown by offshore Santa Ana winds during the 2003 firestorm in southern California, 24 October 2003. These winds occur every autumn after the summer drought with gusts >100 km hr^{-1} and relative humidity <5%. Panel covers an area of ~350 × 600 km; US–Mexico border indicated by thin line about mid-frame (http://earthobservatory.nasa.gov/).

beginning with the 24 to 28 September 1889 Santiago fire (Keeley and Zedler, 2009). These large fire events span the period from before active fire suppression to the present, when active fire suppression is practiced. This illustrates that large wildfires are a natural feature of this landscape and that, despite the best intentions, firefighters are unable to suppress all fires. Although firefighters contain most fires at much smaller sizes than would be the case in the absence of fire suppression activities, the potential still persists for some fires to escape control, particularly under extreme weather conditions.

Although one major aspect of the prior condition is the regional propensity for large, high-intensity wildfires owing to the Mediterranean-type climate and regular Santa Ana wind conditions, the other major condition that contributed to the impact of the 2003 and 2007 wildfires is the distribution of human settlements relative to fire-prone wildland vegetation. The 1889 Santiago fire is estimated to have been the largest recorded fire in the region, yet no one died and no homes were destroyed. Since 1889, however, human population and the area of urban development have grown by orders of magnitude. In the last fifty years, the region has, on average, lost 500 homes a year to wildfires (Cal Fire, 2000). The massive losses of property and lives in recent fires are the result of human population growth and expansion into these fire-prone landscapes.

5.2 Vulnerability

In general, urban environments in southern California are particularly vulnerable to wildfires because of the hot Santa Ana winds, which last several days and have gusts exceeding 100 km/h and relative humidity under 5 per cent. These winds blow from the interior toward the coast, and there are one or more such events every year in the autumn (Raphael, 2003), when vegetation is at its driest. Although the 2003 and 2007 fires were driven by Santa Ana winds, these winds were not outside the normal range of variability in duration or intensity (Keeley et al., 2004; 2009), so it is apparent that winds alone cannot account for why these fires were particularly destructive.

One reason the southern California region was especially vulnerable to massive fire events in 2003 and 2007 is the extraordinarily long antecedent droughts. Annually, the region is subject to an intense summer drought of little or no rainfall for four to six months; however, prior to the 2007 fires, there had been seventeen months of drought with an average Palmer Drought Severity Index (PDSI) of -3.62, and prior to the 2003 fires, there were fifty-four months of drought (Keeley and Zedler, 2009).

Although drought typically affects fire behaviour by decreasing fuel moisture, this was not likely the main reason these droughts contributed to the extraordinary size of these fires. Nearly every autumn, there are Santa Ana wind-driven fires, and the fuel moisture of these shrublands is typically at the lowest level of physiological tolerance. At the time of the 2007 fires, live fuel moisture for the most common chaparral shrub, *Adenostoma fasciculatum*, was no different than in other years (Keeley and Zedler, 2009). It is hypothesised that the primary effect of the drought was to produce significant amounts of dieback in the vegetation, and this contributed to fire spread by increasing the incidence of spot-fires ahead of the fire front (Keeley and Zedler, 2009). This resulted in extraordinarily rapid fire spread that in many cases exceeded firefighters' capacity for defending homes.

5.3 Resilience

The resilience of urban communities to the wildfires of 2003 and 2007 was largely a function of their location and spatial arrangement, as well as the specific properties of home construction and landscaping. Santa Ana wind-driven fires follow specific topographic corridors (Moritz et al., 2010), and at a landscape scale, homes that burned in the 2003 and 2007 fires were distributed in areas that have been historically fire prone and in areas that were located farther inland, closer to the points of fire origin (Syphard et al., 2012). Homes at low to intermediate densities and in smaller, isolated neighbourhoods were also more likely to be burned (Syphard et al., 2012). This could be because of the spatial relationship between homes and wildland vegetation as well as the more limited accessibility of homes to firefighters, as neighbourhoods with fewer roads were also more likely to be burned. Those homes on the interior of developments or on the leeward side (i.e. southern and western perimeters) largely survived untouched (C. J. Fotheringham, US Geological Survey, Western Ecological Research Center, unpublished data).

Homes in the direct path of these fires were the most vulnerable, but housing construction also plays a role. Building ordinances have increased the resilience of homes to burning through structural changes that make homes more resistant to ignition, and as a result, new homes are often more resilient to burning (Quarles et al., 2010). However, another factor is that landscaping age affects the level of plant biomass in close proximity to the homes, and many of these landscaping choices pose significant threats as they age. In a study of 2003 and 2007 fires, the age of homes was significantly correlated with the total tree cover within a 22 m radius of the house ($r^2 = 0.244$, $p < 0.001$, n = 310; C. J. Fotheringham, unpublished data), and as discussed below, this may contribute to structural losses. Thus, it will require some work to parse out the relative role of improved construction techniques from increased landscaping fuels.

5.4 Physical Characteristics of the Event

There is evidence that most of the homes lost in these 2003 and 2007 fires ignited from embers blown from the wildland to the urban environment. An in-depth case study of a neighbourhood that experienced substantial home losses in 2007 showed that two out of every three homes were ignited either directly or indirectly from embers, as opposed to uninterrupted fire spread from the wildland to the structure (Maranghides and Mell, 2010). Another reflection of the importance of embers is the observation that for houses on the perimeter of developments, the amount of clearance around homes had no significant effect on whether a home burned (Table 5.1). These patterns fit a widely held generalisation that most homes are not destroyed by direct heating from the fire front but rather from embers that ignite fine fuels in, on or around the house (Cohen, 2000; Koo et al., 2010). Such embers or firebrands are often carried from fuels several kilometres away, and no reasonable amount of clearance around the home can protect against this threat. The extent to which embers create a hazard is a function of them landing on a suitable fine fuel on or adjacent to the home.

Urban landscaping played a significant role in property losses during the 2003 and 2007 fires (Table 5.1). Homes that burned had significantly greater ground surface

Table 5.1. *Comparison of characteristics of burned and unburned houses in a portion of the 2003 and 2007 fires. Clearance is for a subset of homes on the periphery of urban development. P values for Mann-Whitney test (C. J Fotheringham and J. E. Keeley, unpublished data)*

	Burned		Unburned		
	Mean (S.E.)	N	Mean (S.E.)	N	*p*-value
Clearance width (m)	9.38 (1.27)	83	12.45 (1.41)	82	0.115
Tree canopy overlap (m)	10.79 (1.01)	150	5.37 (0.81)	160	0.00001
Tree ground surface cover (m^2)	146.74 (13.43)	150	97.75 (9.22)	160	0.021
Patio (m)	4.82 (0.45)	150	3.59 (0.35)	160	0.051
Deck windward side (m)	0.87 (0.20)	150	0.383 (0.10)	160	0.069

cover (GSC) of trees in the yard, and the amount of tree canopy that overlapped the house was greater than for unburned homes. It is hypothesised that tree canopies that shade homes drop highly flammable litter on and around the structure that contributes to ignitions from embers. Burned homes also had more patio and decking than unburned homes, and these too potentially contributed to ignition (Table 5.1).

Other landscaping choices that can affect structure loss are the planting of drought-tolerant species by home owners and landscape specialists. Many of these species come from other fire-prone regions and share characteristics with fire-prone native species, including retention of dead fuels in the canopy and increased flammability. In addition, it is commonly believed that ornamental vegetation is resistant to fire because of regular irrigation; however, no systematic studies have been conducted to determine the effects of the extreme Santa Ana conditions on ornamental plant moisture content.

5.5 Emergency Management

In 2007, roughly a half million residents were evacuated, with major disruptions in personal and professional lives. There are numerous ways to calculate the costs of such an event, and estimates range from hundreds of millions to billions of dollars. There are many indirect economic costs that are more difficult to estimate, for example, the displacement of the San Diego Chargers football game to Arizona because of occupation of their stadium by evacuees. Calculating the net economic impact is made even more complicated by the fact that there were huge offsets in the damage from wildfires by insurance payments that amounted to billions of additional dollars to the California economy (Hartwig, 2007).

The massive evacuation from homes in the path of the 2003 and 2007 fires would seem to have been the prudent thing to do, although, despite stern warnings from the media, most agencies involved in this evacuation contend that it was not mandatory. After the 2007 fires, it was advocated that one of the communities that suffered major home losses consider encouraging residents during the next fire to stay with their homes in order to assist firefighters (Paveglio et al., 2010). In southern California, this is referred to as 'shelter-in-place' and is fashioned after a program in Australia known as the 'go early or stay and defend' policy (Mutch et al., 2010). Key to this idea is that it requires pre-planning and decision making long before a fire incident occurs (see further discussion of this policy in Chapter 8).

5.6 Post–Event Adaptation

Understandably, after the enormous impact of the 2003 and 2007 fires, there has been strong public sentiment to try to prevent such losses from occurring again. For example, there was renewed interest in promoting community involvement in fire protection. The federal government has made large sums of money available to community groups such as Fire Safe Councils, whose objectives are to promote fire-safety education for homeowners and to encourage pre-fire management. One of the primary objectives of pre-fire management has been to increase efforts to reduce hazardous fuels. As a result, wildland fuel treatments in southern California U.S. Forest Service (USFS) forests have increased in the years following these 2003 and 2007 fires. In particular, the trend has been to broaden the areal extent of treatments beyond the traditional practice of creating long, linear breaks in vegetation to provide firefighter access for suppression.

Despite these efforts to reduce broad swaths of fuel across the landscape, recent research demonstrates that fuel breaks are most effective where they provide access for fire-fighting activities (Syphard et al., 2011a; 2011b). Therefore, some managers are also starting to recognise that strategically located fuel modification zones around the urban interface are likely to provide better community protection with fewer resource impacts to natural ecosystems (Witter and Taylor, 2005). In addition to strategically located fuel breaks, creating defensible space around homes is now widely embraced in the fire management, policy and scientific communities and strongly promoted in Fire Safe Councils. Defensible space is also likely to be more instrumental in community protection than remotely located fuel breaks.

Despite a legal mandate in California for 30 m clearance around homes, there has been increasing sentiment after the 2003 and 2007 fires that more clearance is always better (e.g. Figure 5.2). Therefore, in many communities, home owners are now requested to clear up to 90 m by the local fire department. In some cases, insurance companies require 120 m. It is increasingly evident from field inspections as well as from aerial imagery that many homeowners at the wildland-urban interface are clearing in excess of 30 m, and some in excess of 90 m.

Figure 5.2 Clearance around a rural home in San Diego County, California, that exceeds state requirements (photo by J. E. Keeley).

A common misunderstanding regarding defensible space is that the words 'vegetation clearance' confuse people into thinking that they need to clear all fuel within the safety zone, that is, to bare ground (e.g. Figure 5.2), instead of simply reducing concentrated fuel around the home. Complete removal of fuel may actually create more problems than it solves; it encourages growth of highly combustible grasses, with a substantially longer fire season; it is aesthetically less pleasing (e.g. Figure 5.2); it degrades the water-holding capacity of the soil, promoting erosion; and it destroys important wildlife habitat essential to birds and small mammals that add to the rural lifestyle (Halsey, 2005).

The current lack of a clear science-based system for determining appropriate clearance size potentially has huge impacts on the landscape. In other words, although the mandate is to create 30 m of defensible space, empirical evidence is still lacking on whether more area will provide more protection. We conducted a rough experiment to estimate approximately how much vegetation removal would occur if defensible space guidelines were strictly adhered to, at both 30 m or increased to 90 m, by residential property owners in San Diego County that owned sufficient land to comply with these guidelines. Using a parcel boundary map and a digitised map of residential structures, we calculated the number of properties that were large enough to accommodate defensible space clearing requirements and multiplied these by the area of the

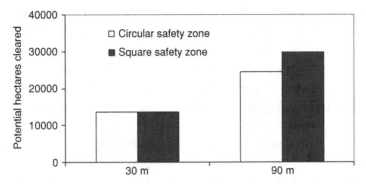

Figure 5.3 Potential area of vegetation clearance that would occur if San Diego County, California, property owners with large-enough properties adhered to 30 m vs. 90 m defensible space requirements. Circular and square safety zones refer to the shape of the clearance around all sides of an occupied structure.

circle that would account for defensible space around all sides of a house located in the centre of the property.

If all property owners on sufficiently sized parcels cleared a 30 m radius around their property, 13,722 ha of vegetation would be removed (Figure 5.3). If those property owners with parcels large enough, that is, a subset of the 30 m parcels, extended clearance to a 90 m radius, this would potentially bring the total vegetation loss to a figure equalling two-thirds the size of the region's largest habitat conservation area, the San Diego Multiple Species Conservation Plan. Of course, many of these parcels already have thinned the vegetation on their property, and not all properties are compliant or will be required to clear the full 90 m around their property. But the numbers do provide a perspective on the importance of understanding the benefits of increasing clearance from 30 to 90 m, which would represent a major loss of natural resources.

5.7 Climate Change Impacts on Southern California Fire Regimes

Southern California is recognised as one of the most fire-prone environments on earth because of its location, climate and vegetation. There is widespread concern that global warming will result in more frequent and more intense fires (Running, 2006). While some landscapes may experience more frequent fires and others more intense fires, it is of course unlikely both will occur in the same ecosystem since they are generally inversely related – that is, intensity is heavily dependent on duration of fuel accumulation.

It is our view that most of the published forecasts of climate change impacts on fire regimes are rather speculative at this point. Predictions are largely based on increasing temperatures affecting fire activity by reducing fuel moisture, which often is tied to increased probability of ignitions and fire spread. There are several considerations that need to be looked at before accepting this causal relationship. (1) Global warming is driven by increased partial pressure of CO_2, and there are direct effects of increased CO_2 on plant physiology that will act to produce opposite effects on fuel moisture. In short, as CO_2 goes up, water use efficiency goes up, and for chaparral, this has been estimated to be as much as 35 per cent with a doubling of CO_2 (Chang, 2003). (2) Fire regimes of different vegetation types will not likely respond the same to increased temperatures and increased CO_2, and since fire regime changes have the potential for type conversion of vegetation, predictions of future fire activity cannot be made without serious consideration of vegetation changes. (3) Climate change is only one of a multitude of global changes. In southern California, models predict a 3 to 5 per cent increase in temperature but more than a 50 per cent increase in population by 2040. Since humans are responsible for more than 95 per cent of all fire ignitions, and expansion of urban development into wildlands sets the stage for catastrophic wildfire outcomes, predictions about future fire impacts that fail to include human demographic changes are of questionable value for this region.

Changes in winds have the potential for substantial changes in future fire regimes, but we have even less certainty as to what to expect with winds. Some models of future changes in Santa Ana winds suggest a shift to later in the autumn, and Miller and Schlegel (2006) predict that this will result in increased area burned in coastal California. However, one could predict the opposite effect because later winds will increase the probability of Santa Ana winds being preceded by autumn rains, and historically, when winds have been preceded by precipitation, it has had a negative effect on area burned (Keeley, 2004). In a different modelling framework, Hughes, Hall and Kim (2009) predicted a dramatic drop in Santa Ana winds in the coming years, which of course would suggest we have a rosy future in terms of reduced fire hazard. In short, the models we have for predicting the future are often contradictory, which is to be expected because they are in a rudimentary stage of development. However, as a consequence, they are not presently useful for most of the decision making required to deal with future fire hazards in the region.

5.8 What Are the Lessons Learned?

The 2003 and 2007 wildfires remind us that large fire events are an inevitable and inescapable part of living in southern California. Despite decades of fuel break construction, improvements in fire-safe codes and building regulations and thousands of firefighters, homes continue to be lost nearly every year. The predominant viewpoint has been that government is responsible for protecting homes during fires; but as scattered patterns of development continue to extend into the most flammable parts of the landscape, it becomes more and more difficult for firefighters to defend every home. Thus, many victims blame government officials for not having cleared more fuel, or they accuse firefighters of not protecting their homes (Kumagai et al., 2004).

Perhaps because most fire management has been focused on wildlands, there has been relatively little effort towards learning from other hazard sciences. For example, flood hazard science has made great strides in reducing losses through better land planning (Abt et al., 1989). The potential is immense for fire scientists and emergency mangers to learn from these hazard sciences, as altered land planning is very likely one of the more important avenues for reducing losses from wildfires as well. This is because the location and pattern of housing significantly influence where fires occur and, in turn, where fires are most likely to result in losses.

Earthquake science has never taken the approach of trying to eliminate the hazard but rather alters human infrastructure to make living in this environment much safer. Fire scientists are gradually coming around to the idea of infrastructure hardening, but most information on the types of construction and landscaping necessary to fire-proof a house are of an anecdotal nature, and there is an urgent need for science-based approaches. We suggest a change in perspective that acknowledges fire risk as an inevitable component of the landscape and that we prepare as we would for other hazards.

References

Abt, S. R., Witter, R. J., Taylor, A. and Love, D. J. (1989). Human stability in a high flood hazard zone. *Journal of the American Water Resources Association*, 25, 881–890.

Cal Fire (2000). *Wildland Fire Hazard Assessment. Final Report on FEMA 1005–47.* Sacramento: California Division of Forestry and Fire Protection.

Chang, Y. (2003). *Effects of Manipulated Atmospheric Carbon Dioxide Concentrations on Carbon Dioxide and Water Vapor Fluxes in Southern California Chaparral.* Unpublished PhD thesis. Davis: University of California, and San Diego: San Diego State University.

COES – California Office of Emergency Services (2004). *2003 Southern California Fires. After Action Report.* Sacramento: California Office of Emergency Services, Planning and Technological Assistance Branch.

Cohen, J. D. (2000). Preventing disaster: home ignitability in the wildland-urban interface. *Journal of Forestry*, 98, 15–21.

Halsey, R.W. (2005). *Fire, Chaparral, and Survival in Southern California.* San Diego, CA: Sunbelt Publications.

Hartwig, R. P. (2007). *The California Wildfire Season of 2007.* Washington, DC: Insurance Information Institute.

Hughes, M., Hall, A. and Kim, J. (2009). *Anthropogenic Reduction of Santa Ana Winds. CEC-500–2009–015-D.* Sacramento: California Climate Change Center.

Keeley, J. E. (2004). Impact of antecedent climate on fire regimes in coastal California. *International Journal of Wildland Fire*, 13, 173–182.

Keeley, J. E., Fotheringham, C. J. and Morais, M. (1999). Reexamining fire suppression impacts on brushland fire regimes. *Science*, 284, 1829–1832.

Keeley, J. E., Fotheringham, C. J. and Moritz, M. A. (2004). Lessons from the 2003 wildfires in southern California. *Journal of Forestry*, 102, 26–31.

Keeley, J. E., Safford, H., Fotheringham, C. J., Franklin, J. and Moritz, M. (2009). The 2007 southern California wildfires: lessons in complexity. *Journal of Forestry*, 107, 287–96.

Keeley, J. E. and Zedler, P. H. (2009). Large, high intensity fire events in southern California shrublands: debunking the fine-grained age-patch model. *Ecological Applications*, 19, 69–94.

Koo, E., Pagni, P. J., Weise, D. R. and Woycheese, J. P. (2010). Firebrands and spotting ignition in large-scale fires. *International Journal of Wildland Fire*, 19, 818–843.

Kumagai, Y., Bliss, J. C., Daniels, S. E. and Carroll, M. S. (2004). Research on causal attribution of wildfire: an exploratory multiple-methods approach. *Society and Natural Resources*, 17, 113–127.

Maranghides, A. and Mell, W. (2010). A case study of a community affected by the Witch and Guejito wildland fires. *Fire Technology*, 47, 379–420.

Miller, N. L. and Schlegel, N. J. (2006). Climate change projected fire weather sensitivity: California Santa Ana wind occurrence. *Geophysical Research Letters*, 33, L15711, doi:10.1029/2006GL025808.

Moritz, M. A., Moody, T. J., Krawchuk, M. A., Hughes, M. and Hall, A. (2010). Spatial variation in extreme winds predicts large wildfire locations in chaparral ecosystems. *Geophysical Research Letters*, 37, L04801, doi:10.1029/2009GL041735.

Mutch, R. W., Rogers, M. J., Stephens, S. L. and Gill, A. M. (2010). Protecting lives and property in the wildland-urban interface: communities in Montana and Southern California adopt Australian paradigm. *Fire Technology*, 47, 357–377.

Paveglio, M., Carroll, S. and Jakes, P. J. (2010). Adoption and perceptions of shelter-in-place in California's Rancho Santa Fe fire protection district. *International Journal of Wildland Fire*, 19, 677–688.

Quarles, S. L., Valachovic, Y., Nakamura, G. M., Nader, G. A. and De LaSaux, M. J. (2010). *Home Survival in Wildfire-prone Areas: Building Materials and Design Considerations.* Richmond: University of California, Agriculture and Natural Resources.

Raphael, M. N. (2003). The Santa Ana winds of California. *Earth Interactions*, 7, 1–13.

Running, S. W. (2006). Is global warming causing more, larger wildfires? *Science*, 313, 827–928.

Syphard, A. D., Keeley, J. E. and Brennan, T. J. (2011a). Comparing the role of fuel breaks across southern California national forests. *Forest Ecology and Management*, 261, 2038–2048.

Syphard, A. D., Keeley, J. E. and Brennan, T. J. (2011b). Factors affecting fuel break effectiveness in the control of large fires in the Los Padres National Forest, California. *International Journal of Wildland Fire*, 20, 764–775.

Syphard, A. D., Keeley, J. E., Massada, A. V., Brennan, T. J. and Radeloff, V. C. (2012). Housing arrangement and location increase wildfire risk. *PLoS ONE*, 7(3), e33954, doi:10.1371/journal.pone.0033954.

Witter, M. and Taylor, R. (2005). Preserving the future: a case study in fire management and conservation from the Santa Monica Mountains. In *Fire, Chaparral, and Survival in Southern California*, ed. R. W. Halsey. San Diego, CA: Sunbelt Publications, pp. 109–15.

6

Adapting to Extreme Heat Events: Thirty Years of Lessons Learned from the Kansas City, Missouri, Extreme Heat Program

DAVID M. MILLS AND WILLIAM D. SNOOK

Extreme heat events (EHEs) have been defined as 'summertime weather that is substantially hotter and/or more humid than average for a location at that time of year' (USEPA, 2006). Each year, EHEs adversely affect human health in locations around the world. Severe EHEs have resulted in hundreds to tens of thousands of deaths (e.g. Chicago in 1995 and Western Europe in 2003). These losses are all the more tragic considering most public health officials now view these deaths as preventable (e.g. Centers for Disease Control and Prevention, 2009; World Health Organization, 2009).

EHEs are of particular interest when assessing climate change impacts and adaptation for three main reasons. First, because EHEs are often identified using meteorological thresholds, researchers can quantify anticipated increases in the frequency and severity of future EHEs (e.g. Meehl and Tebaldi, 2004; Diffenbaugh and Ashfaq, 2010). Second, because health response functions can be developed for EHE conditions (e.g. Medina-Ramon and Schwartz, 2007; Anderson and Bell, 2009), anticipated changes in EHE-attributable health impacts can be quantified (e.g. Peng et al., 2011). Finally, the experience of extreme heat programs (EHPs) provides valuable insight and lessons learned for other climate change adaptation efforts.

Kansas City, Missouri's EHP was developed in response to a severe EHE in July 1980. A review of its evolution and lessons learned over thirty years of operation provides an opportunity to inform other climate change adaptation efforts.

This chapter proceeds as follows: Section 6.1 provides an overview of conditions and impacts of the July 1980 EHE in Kansas City; section 6.2 describes critical events in the evolution of Kansas City's EHP; section 6.3 summarises key lessons learned; and section 6.4 provides concluding remarks.

6.1 July 1980 EHE in Kansas City

The summer of 1980 was extremely hot in the United States. In Kansas City, it was one of the ten hottest summers on record and the hottest since the early 1950s

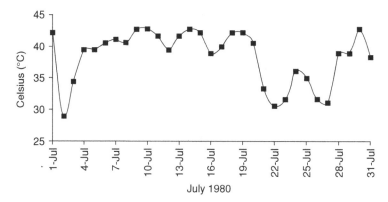

Figure 6.1 Maximum daily temperatures in downtown Kansas City, Missouri, during July 1980.

(National Weather Service, 2010). Figure 6.1 presents the daily maximum downtown temperatures for July when the heat peaked. From 4 July through 20 July, daily maximum temperatures never fell below 38.9 °C (102 °F), while daily minimum temperatures remained above 26.6 °C (79.9 °F). As a result, the population received little cooling relief (Jones et al., 1980).

This heat caused a dramatic increase in the number of deaths in Kansas City, Missouri. In July 1980, 598 deaths were recorded in contrast to 362 deaths in July 1979. This suggests the EHE contributed to a 65 per cent increase in mortality (Jones et al., 1982). This research also showed that health impacts were disproportionate in a number of population subgroups, including the poor, elderly and non-whites (Jones et al., 1982).

6.2 Kansas City's EHP

Because Kansas City had experienced prior EHEs, several response actions were implemented in July 1980 to minimise EHE-attributable health impacts. For example, fans were distributed to families and a phone-based 'heat relief hotline'[1] was established for residents to request assistance and receive information (Jones et al., 1980). In addition, the severity and magnitude of the health impacts attributable to the 1980 EHE spurred the city to develop an EHP as part of its all-hazards emergency operations response plan.

Following the 1980 EHE, Kansas City established a Heat Task Force and the city's Health Department was assigned with coordinating EHE identification, notification, and response efforts. From 1980 to 2006, the Heat Task Force focused solely on Kansas City, Missouri, proper, not the larger metropolitan area. Activities included distributing

---

[1] In the 1980s, the hotline was staffed by health department employees during EHEs. Hotline staff listen to the caller's needs/requests and then either provide additional information or help coordinate the services required to address health issues (e.g. home health check by medical personnel).

fans to vulnerable residents through 'fan clubs', installing air conditioners for at-risk residents and increasing reporting on heat-related fatalities within the medical system. In addition, the Health Department began to issue heat-health notifications (e.g. advisories, warnings, emergencies) using criteria for identifying future EHEs based on health-impact data gathered in 1980.

In 2007, Kansas City's EHP changed significantly in an effort to minimise EHE health impacts and better coordinate program resources by bringing together federal, state and local partners to develop a region-wide approach to addressing EHEs. With the addition of new partners, the city's Health Department continued to serve as the lead agency responsible for the city's EHP and expanded its role to be the central point of communication in the new partnership. Within the revised program, the Health Department's main responsibilities included:

- assisting the National Weather Service in determining whether to issue heat-health notifications;
- notifying program partners with respect to EHE decisions;
- coordinating EHE-related response activities;
- developing and coordinating public messaging, media and communication strategies for EHE risks;
- collecting and summarising information on the health impacts and EHE response measures for specific events throughout the summer;
- suggesting improvements to health surveillance systems; and
- coordinating pre- and end-of-season meetings with partners to review community plans and evaluate performance.

Figure 6.2 summarises the nature and flow of information in this partnership, with the Health Department serving as the program's information clearinghouse.

The 2007 program changes stemmed from a belief that those most vulnerable to EHEs could be more effectively accessed, evaluated and assisted by engaging those most experienced in working with these high-risk individuals. The 2007 program revisions were designed to better utilise existing 'messengers' for communicating the program's objectives and educational messages to the general public and specific high-risk population subgroups. For example, the Salvation Army and the United Way, which operates the region's 2–1–1 social service call centre, became critical partners for coordinating and delivering services during EHEs. It is widely believed that incorporation of these and other experienced groups have helped to improve the program's performance.

As a result of these changes, the EHP now focuses on three elements to protect Kansas City metro residents:

- regional resource coordination (e.g. distributing information on cooling shelter locations, operating call centres);
- identification and assessment of EHE response (e.g. temperatures, gaps in service or information, at-risk individuals); and

Active Heat Communication and Surveillance

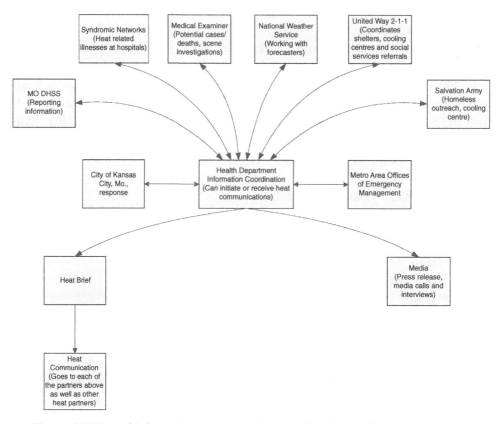

Figure 6.2 Flow of information among the Kansas City Health Department and its EHP partners.

- media relations – assessing and evaluating the adequacy of public EHE/EHP information and messages and the interface of EHE/EHP messaging with and use of different media resources to distribute these messages.

These elements exhibit the city's belief that it is not enough for EHP partners to merely provide the public with information and hope they respond as anticipated and adopt appropriate behaviours. Partners must actively seek out those at greatest risk, assess their health status and, when necessary, intervene to provide relief. This approach recognises that the characteristics that define high-risk subpopulations (e.g. advanced age, poverty, physical or cognitive limitations) also typically limit their ability to access and act upon heat safety messages. For example, the Salvation Army has extensive local experience providing services to the area's homeless through efforts such as providing meals. During a heat warning, the Salvation Army draws on its knowledge to dispatch its service trucks with cool water to daytime locations where the homeless and others may easily access the service (see Figure 6.3).

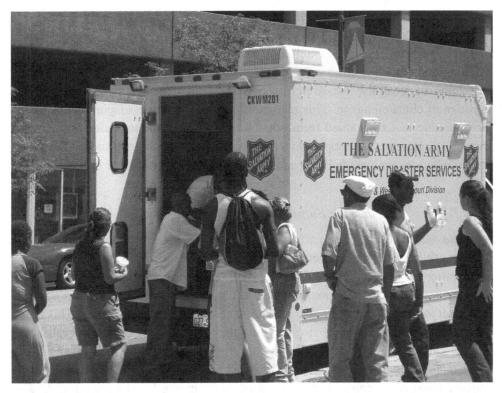

Figure 6.3 The Salvation Army distributing bottled water during a Kansas City EHE (courtesy of the Salvation Army).

Critical to the EHP's continued success are pre- and end-of-season meetings to review needs and evaluate performance. These meetings also try to identify and recruit additional partners to help address high-risk subgroups that remain difficult to access and serve (e.g. the elderly, 'shut-ins', the homeless). The consistent evaluation of information needs and program performance also ensures a dynamic EHP structure that guarantees a quick response to changing situations. These meetings also recognise that a more formal program evaluation effort is not within the program's scope because of data and resource constraints. Despite this lack of a formal evaluation, compelling evidence of the program's success is available from the individuals in distress that have been identified and assisted by program partners during EHP-related activities, helping to avoid a more severe heat-attributable adverse health outcome.

6.3 Lessons Learned for Climate Change Adaptation Efforts

Over its thirty years of operation, Kansas City's EHP managers have learned a number of important lessons relevant to climate change adaptation efforts. This section briefly summarises these lessons.

Reinforce vs. Reinvent Responses

EHPs and climate change adaptation efforts can draw on a range of current hazard planning and response efforts. A critical element of climate change adaptation is the ability to recognise the similarity between current health stressors and those sensitive to climate change. This recognition should improve adaptation efforts by incorporating lessons learned from other similar efforts. In short, although climate change will increase the stress from a number of direct and indirect sources, many of these challenges are already being addressed in other contexts.

Place the Future in the Appropriate Context

As noted, Kansas City's EHP partners believe regular planning and evaluation activities are critical program components. Planning involves being aware of and actively considering information that could affect EHE conditions and responses in the short term and in the long term. Currently, Kansas City's EHP managers and partners find it difficult to actively plan for anticipated climate-driven changes in EHEs. This is not a rejection of available climate change information. Rather, it is a practical realisation that the composition and location of at-risk groups, as well as the urban landscape, are changing rapidly. Also, program partners and agencies work together to maximise existing resources because there is no dedicated funding for the program. At any point, partners may need to scale back participation if funding for their day-to-day programs is reduced or eliminated. Because these changes require significant program adjustments, integrating the relatively uncertain nature and timing of climate change impacts on EHEs remains beyond the scope of current program planning and response activities. This suggests a need for climate researchers to better understand the types of information needed by those developing adaptation efforts in order to more effectively prepare for the potential impacts of climate change.

Always Try to Improve Public Education, but Be Ready to Act

Kansas City's EHP continues to actively focus on educating the public about the risks and appropriate responses to EHEs. This requires identifying ways to publicly convey EHE messages – for example, by posting National Weather Service heat warnings during regular television programming and recruiting new partners, such as non-English-speaking media, to relay messages to their respective audiences. By most accounts, these efforts have increased the overall level of public awareness and acceptance of EHEs as a significant public health threat in Kansas City. However, there is limited evidence that most of the population changes their behaviour during an EHE (e.g. Sheridan, 2007). This suggests that continued public education is needed because EHEs are likely to become more frequent and severe and characteristics of vulnerable groups are likely to change over time.

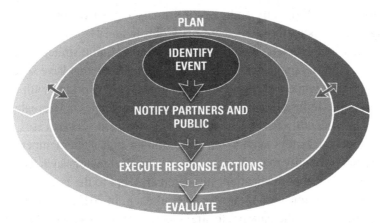

Figure 6.4 Elements for a successful EHP or response to a climate-sensitive stressor. Event-specific actions appear in the centre figure with the overlapping spheres of activity; longer-term supporting efforts are represented by the interacting elements of the outer ring.

EHP partners are acutely aware that many of the characteristics elevating the risk of experiencing an EHE-attributable adverse health outcome may preclude receiving or effectively acting on the EHE information provided. The partners recognise that their efforts to directly assess and assist those at greatest risk will remain a critical component of the EHP. The ability to form partnerships that are ready and able to directly address a climate-sensitive stressor is a key consideration in a range of climate change adaptation planning contexts.

There Are Many Useful Tools, but No One Tool

Kansas City's EHP relies on partners to use the tools and methods they have developed to assess and serve local populations and also to provide a consistent educational message about EHEs. As a result, the program's perceived success is not clearly attributable to a single set of actions but to the combined effort of all partners. Figure 6.4 provides a summary of how different efforts within Kansas City's EHP relate to event-specific actions, the inner spheres of activity and longer-term program support activities (the outer ring of actions). The relationships among the elements in Figure 6.4, although presented here for EHPs, are applicable to a wide range of climate change adaptation efforts.

Partnerships Enhance Adaptation Opportunities

By actively developing partnerships for its EHP, the Kansas City Health Department has greatly enhanced the potential to meet the program's public health objectives. Specifically, by drawing on partners with specialised resources, information and relationships, the EHP directly addresses a primary challenge for any intervention: Identifying and assessing those at greatest risk during an EHE.

Other climate adaptation efforts could benefit from a similar strategy, that is, one in which adaptation program leaders collaborate with partners that have specific knowledge or experience while pursuing a common goal. While working with the Jackson County, Missouri, Medical Examiner's Office, the two agencies have created a preliminary heat-related death reporting form for suspected heat-related cases (Appendix 6.1). Environmental factors such as air-conditioning unit status temperature at the location and fan proximity to body in addition to medical factors such as alcohol use, mental health and current medications are noted and sent to the health department to aid in tailoring communication messages and outreach. In 2012, having the preliminary heat reports allowed the partners to stress messages on mental health, alcohol use and being outside in a car or on hot concrete, as only three of the ten deaths occurred in a home environment.

Additionally, during the 2011 summer, Kansas City had thirty-four days with a heat advisory or excessive heat warning. Through outreach and education efforts with local broadcasters, program partners were able to document 821 separate media 'hits', or mentions, in television broadcasts, which were a primary vehicle for providing educational messages to the public. Using the Nielsen data of estimated television audiences, provided by the Health Department's contracted media tracking service, it is estimated those hits were associated with more than to 33.4 million possible media impressions (people who may have seen the messages). Through an aggressive communication campaign, EHE partners were able to identify residents or special-needs populations who needed 'welfare checks' that were coordinated through local service organisations and law enforcement.

6.4 Conclusions

Although Kansas City has made advances in addressing the public health threat of EHEs, considerable challenges remain. Climate change will increase the risk posed by future EHEs by increasing their frequency, severity and duration. Simultaneously, those most vulnerable during EHEs are likely to continue to find it difficult to effectively act upon critical public education messages regarding EHEs.

Kansas City's approach provides a model for other adaptation efforts. Specifically, a climate adaptation program can improve its capabilities through active partnerships. Additionally, active program planning and evaluation will keep the effort vibrant and responsive to changing conditions. Finally, continuous and consistent public education is critical for reinforcing desired messages. However, partners must be willing to take direct action to minimise adverse impacts.

Kansas City's experience also identifies a need for climate change researchers to more actively engage those working on adaptation programs in order to better understand information they need to address climate change in current planning efforts. Specifically, climate researchers seeking to improve the policy relevance of their results should consult with program managers to understand the spatial resolution, timing and levels of certainty that are required for results to be considered as well as what would be most desirable.

References

Anderson, B. G. and Bell, M. L. (2009). Weather-related mortality: how heat, cold and heat waves affect mortality in the United States. *Epidemiology*, 20, 205–213.

Centers for Disease Control and Prevention (2009). *Heat Waves*. Atlanta, GA: Centers for Disease Control and Prevention. Accessed 23 October 2012 from: <http://www.cdc.gov/climateandhealth/effects/heat.htm>.

Diffenbaugh, N. S. and Ashfaq, M. (2010). Intensification of hot extremes in the United States. *Geophysical Research Letters*, 37, 1–5.

Jones, S., Griffin, M., Liang, A. and Patriarca, P. (1980). *The Kansas City Heat Wave, July 1980: Effects of Health, Preliminary Report*. Atlanta, GA: Centers for Disease Control.

Jones, S. T., Liang, A. P., Kilbourne, E. M. et al. (1982). Morbidity and mortality associated with the July 1980 heat wave in St. Louis and Kansas City, MO. *Journal of the American Medical Association*, 247, 3327–3331.

Medina-Ramon, M. and Schwartz, J. (2007). Temperature, temperature extremes, and mortality: a study of acclimatisation and effect modification in 50 US cities. *Occupational and Environmental Medicine*, 64, 827–833.

Meehl, G. A. and Tebaldi, C. (2004). More intense, more frequent, and longer lasting heat waves in the 21st century. *Science*, 305, 994–997.

National Weather Service (2010). *Kansas City/Pleasant Hill, MO – Summer Ranking by Average Temperature 1889–2011*. Pleasant Hill, MO: National Weather Service Weather Forecast Office. Accessed 17 October 2012 from: <http://www.crh.noaa.gov/eax/localclimate/seasrank/summertrank.php>.

Peng, R. D., Bobb, J. F., Tebaldi, C. et al. (2011). Toward a quantitative estimate of future heat wave mortality under global climate change. *Environmental Health Perspectives*, 119, 701–706.

Sheridan, S. C. (2007). A survey of public perception and response to heat warnings across four North American cities: an evaluation of municipal effectiveness. *International Journal of Biometeorology*, 52, 3–15.

USEPA – United States Environment Protection Agency (2006). *Excessive Heat Events Guidebook. 430-B-06–005*. Washington, DC: United States Environment Protection Agency.

World Health Organization (2009). *Improving Public Health Responses to Extreme Weather/Heat-waves: EuroHEAT. Technical Summary*. Copenhagen, Denmark: World Health Organization Regional Office for Europe.

OFFICE OF THE
JACKSON COUNTY MEDICAL EXAMINER

(816) 881-6600
FAX (816) 404-1345

660 East Twenty Fourth Street
Kansas City, Missouri 64108

Mary H. Dudley, M.D.
Chief Medical Examiner

PRELIMINARY HEAT DEATH REPORTING FORM

This form contains information intended only to notify the appropriate Public Health Agency (PHA) of a potential heat-related death. The death described is not confirmed, and the case is an open investigation; thus, the specific case information, except for year of birth and sex, is not for release until the case is closed (complete).

M.E. CASE # (for case tracking purposes only): _____ **PROCEDURE:** Autopsy / Inv External / MD External / SO

AGE OF DECEDENT: _____ years / months (circle one) **BIRTH YEAR**_____ **SEX**: male / female

ADDRESS (confidential information; used to determine PHA jurisdiction): _____

_____

_____

DATE/TIME OF DEATH: _____ pronounced / found dead (circle one)

DATE/TIME LAST SEEN ALIVE : _____

   Decomposition? yes / no (circle one) (decomposition correlates with longer toxicology TAT)

TYPE OF LOCATION : apartment / house / attic / mobile home / car / outdoors / other (circle one)

Describe type of outdoors or other : _____

   Make of structure if indoors: Brick / Wood / Concrete / other _____

If decedent died in hospital, was an admit temperature documented? Yes / No / NA (circle one)
      If yes, what was the admit temperature? _____F

TEMPERATURE/HUMIDITY:
   AT TIME FOUND OR PRONOUNCED: Outside: Ambient_____F; Surface _____ F
                                        Inside _____F (ambient)
                                        Humidity _____%

   IF FOUND DEAD, FOR THE DATES SINCE LAST SEEN ALIVE, THE TEMPERATURE AND HUMIDITY RANGES FOR THAT ZIP CODE WERE:
      Outside Temperature _____-_____F
      Humidity _____-_____%
   (Previous and current temperature and humidity information by zip code available from www.wunderground.com)

ENVIRONMENTAL FACTORS: (circle one per item)

      Fan: on / off / broken / NA   Approximate distance from fan: _____

      Air conditioner: on / off / broken / NA   Approximate distance from AC: _____

      Windows: open / closed / NA   Approximate distance from window: _____

OTHER SIGNIFICANT FACTORS: (check all that apply)

    MEDICAL HISTORY

        ☐ **Cardiovascular disease**

           ☐ Hypertension   ☐ Coronary artery disease ☐ Atherosclerosis/hyperlipidemia

           ☐ Previous myocardial infarction   ☐ CABG or stent placement ☐ CHF

        ☐ **Pulmonary disease**

           ☐ COPD ☐ Asthma ☐ Emphysema

        ☐ **Obesity** (Height _____ inches; Weight _____ pounds)

        ☐ **Diabetes mellitus**

        ☐ **Neurologic Diseases**

           ☐ Schizophrenia ☐ Alzheimer's disease or other dementia ☐ Epilepsy/seizures

           ☐ Other mental health or neurological problems _____

        ☐ **Taking prescription or over the-counter medications** _____

        ☐ **Other** _____

        _____

    SOCIAL HISTORY

        ☐ **Chronic/previous drug abuse**

           ☐ Alcohol ☐ Methamphetamine ☐ Cocaine/Crack ☐ Other _____

        ☐ **Known or presumed current/recent drug abuse:** ☐ Same as chronic

           ☐ Alcohol ☐ Methamphetamine ☐ Cocaine/Crack ☐ Other _____

        ☐ **Other** _____

        _____

    CIRCUMSTANTIAL FACTORS

        ☐ **Exercising outside**

        ☐ **Working outside**

           ☐ On the job ☐ At home

        ☐ **Inappropriately clothed; Describe:**_____

        ☐ **Lived alone**

        ☐ **Suspicious or criminal case**

        ☐ **Other**_____

        _____

Preliminary Cause of Death (pending additional studies): _____

Investigator Signature: _____ **Medical Examiner Signature**: _____

Date Sent to Health Dept: _____

Medications known to increase risk of heat-related death include neuroleptics (antipsychotics and major tranquilizers), anticholinergics (tricyclic antidepressants, antihistamines, some anti-Parkinsonian meds, and some over-the-counter sleep meds), and diuretics

Part II

Case Studies from Australia

7

Drought and Water in the Murray-Darling basin:
From Disaster Policy to Adaptation

LINDA C. BOTTERILL AND STEPHEN DOVERS

In 1989, Australian government policy ceased to consider drought as a natural disaster but rather viewed it as a problem of risk management in a highly variable climate. This fundamental shift was a long time in the making and has not been without implementation difficulties; however, the experience contains lessons for climate change adaptation for communities and policymakers. Adaptation and disasters are about coping with greater frequency and/or intensity of climate-related events – storms, flooding, heat waves, cyclones and so on. If these events become more norm than exception, then a similar policy shift from disaster preparation and relief to risk management can be expected. The shift is consistent with evolution of emergency and disaster theory and practice, from natural disasters and 'acts of God' to risk management to deal with the intersection of environmental variation and human vulnerability across hazard types (Handmer and Dovers, 2013). Like other insights from cognate policy and research domains, this shift has yet to be comprehended in much adaptation literature (Dovers and Hezri, 2010).

We summarise the history of drought policy in Australia and the political, cultural and other issues embedded in the process of policy change. We then comment on the intersection of drought policy with the closely related issue of water allocation in the Murray-Darling Basin under conditions of scarcity and climate variability.

7.1 Drought Policy Pre–1989

Until 1989, Australian governments regarded droughts as natural disasters and responded in line with their treatment of cyclones, wildfires, earthquakes and floods. Constitutionally, disaster response is a state (provincial) government responsibility. However, the national government increasingly became involved in delivery of disaster relief after its first intervention to assist the State of Tasmania following major wildfires in 1939. Over the next three decades, a cost-sharing arrangement emerged between the

Commonwealth and states, augmented on a number of occasions by special Commonwealth legislation for a number of disasters, including the drought of 1965/66, and was formalised in 1971 as the natural disaster relief arrangements (NDRA). The NDRA sets out the respective responsibilities of the Commonwealth and states in the event of a disaster and provides a formula for cost sharing. Responsibility for declaring a natural disaster remains with the states, and the Commonwealth accepts these declarations as the basis for triggering the activation of the NDRA.

In 1989, the Commonwealth Government decided that drought was no longer to be covered by the NDRA for three reasons. First, throughout 1989, the media carried reports of apparent misuse of drought assistance for party political purposes by the Queensland State Government. Second and related to this were financial considerations. Drought relief was dominating NDRA expenditure, particularly in Queensland, where it accounted for all Commonwealth expenditure on disaster relief in 1984 to 1984/85, 1986/87, and 1987/88 (Courtice, 1989: 1379). Third, scientific advances in understanding of the Australian climate meant it was increasingly untenable to sustain the position that drought was a disaster or 'act of God'. There was growing acceptance that drought was a normal consequence of Australia's highly variable climate.

7.2 The Policy Transition

The removal of drought from the NDRA was a watershed decision. In May 1990, the Minister for Primary Industries and Energy established a Drought Policy Review Task Force to identify policy options that encouraged primary producers and other segments of rural Australia to adopt self-reliant approaches to the management of drought; consider the integration of drought policy with other policies; and advise on priorities for Commonwealth Government action in minimising the effects of drought in the rural sector (Drought Policy Review Task Force, 1990). The Task Force concluded that drought was a normal part of the Australian climate and a risk to be managed by agricultural producers along with the other risks facing farms. It concluded that 'Managing for drought is about managing for the risks involved in carrying out an agricultural business, given the variability of climate. Drought represents the continuing risk that seasonal conditions will not be adequate to sustain agricultural activity' (Drought Policy Review Task Force, 1990: 3).

The Task Force called for development of a national drought policy as a 'matter of urgency' (Drought Policy Review Task Force, 1990: 21). Over the next two years, Commonwealth and state governments negotiated the contents of such a policy. The new National Drought Policy was announced in July 1992, based on principles of self-reliance, risk management and acceptance that drought is a natural feature of the Australian climate (ACANZ, 1992). The objectives of the policy were to:

- encourage primary producers and other sections of rural Australia to adopt self-reliant approaches to managing for climatic variability;
- facilitate the maintenance and protection of Australia's agricultural and environmental resource base during periods of increasing climate stress; and
- facilitate the early recovery of agricultural and rural industries, consistent with long-term sustainable levels (ACANZ, 1992).

The policy was built on a number of Commonwealth programs for farmers that had been reviewed concurrently. The package of measures came into effect on 1 January 1993. The key features of the drought policy were as follows. First, the main direct form of support for farm businesses was delivered through the Rural Adjustment Scheme, which provided subsidies on interest paid on commercial finance ('interest subsidies'). These were available to farmers with a long-term productive future in agriculture and were offered to support measures by farmers to improve farm profitability and sustainability. During severe droughts – 'exceptional circumstances' – the level of subsidy was increased. Second, the Commonwealth offered tax-effective income-smoothing mechanisms in the form of income equalisation deposits and farm management bonds, which allowed farmers to put aside money in high-income years to be used during drought. These two schemes were later combined to form the Farm Management Deposits Scheme, which remains in place. The third major component was the Farm Household Support Scheme. This was aimed at farmers who did not have a long-term sustainable future in agriculture and was linked to exit grants. It provided a level of welfare support but was structured in such a way as to ensure that it did not undermine the policy objective of only supporting farms with a sustainable future. The Farm Household Support Scheme was time limited and contained a loan component as incentive for non-profitable farmers to leave the industry.

In 1994, following political pressure during drought, a major change was introduced with the announcement of the drought relief payment. This was a welfare payment available to all farmers in an exceptional circumstances area, irrespective of the health of their businesses. This was a major policy shift, as previous drought relief had been tightly targeted at those that were 'profitable in the long term' (ACANZ, 1992: 14). Over the next two decades, this payment, renamed the Exceptional Circumstances Payment, became increasingly generous (Botterill, 2006) and began to dominate the drought relief budget. The rhetoric of self-reliance and risk management continued to be employed by both sides of politics, but implementation of the policy became increasingly inconsistent with these objectives. A major problem with the policy implementation relates to the distinction between normal and 'exceptional circumstances' drought. There was no definition in the legislation or in accompanying material that gave effect to the policy. As a consequence, the definition has been subject to ongoing debate. This has undermined the intention of the policy to provide an ongoing, predictable policy response to drought, unlike the *ad hoc* approaches of the pre-1989 policy. For a more detailed account of the development and evolution of Australia's National Drought Policy, see Botterill (2003; 2005). A recent report by Australia's Productivity Commission has recommended that the exceptional circumstances declaration process and the associated government support packages be terminated (Productivity Commission, 2009). An Intergovernmental Agreement on Drought Policy Reform was signed in May 2013, the language of which is strikingly similar to that of the 1992 National Drought Policy. It sets out to 'replace the exceptional circumstances arrangements' with 'a new package of national drought programs' however the focus and structure of these new programs appears little different from their predecessors.

It is worth noting the narrow focus of Australian drought policy on agricultural production. The policy was developed within the agriculture department, then the Department of Primary Industries and Energy, and from 1996 until 2007 the Cabinet Minister responsible for this portfolio was from the farmer-based National Party. This farmer-centric approach meant the impact of drought on town water supplies and indeed on irrigated agricultural production was given scant attention. Some measures to assist farm-dependent small businesses in rural areas were added some years into the program, but predominantly drought policy has been agricultural industry policy. In spite of this narrow focus, the policy can be seen as a pilot study of attitudinal change and adaptation as policy discourse shifted from disaster response to risk management. Although there has been some backsliding from the policy's original objectives, there has been broad acceptance across the political spectrum that drought is not a disaster. In developing its response to a series of inquiries into drought policy in 2008 (Drought Policy Review Expert Social Panel, 2008; Hennessy et al., 2008; Productivity Commission, 2009), the Australian Government in cooperation with the Western Australian State Government ran a drought 'pilot'. A review panel of the progress of the pilot was largely favorable (Drought Pilot Review Panel, 2011).

This short history demonstrates the often long and variable process of policy reform and two key issues of relevance to climate adaptation and extreme events. First, it is likely that, should extremes other than drought – for example, storms, coastal surges, heatwaves – become more frequent, a similar shift from exception to expected may occur, with demands for similar policy shifts. Second, this case evidences the fact that, very often, a considered policy shift can be reconsidered, at least in part. This appears to be the case with, for example, the treatment of the welfare component of the Western Australian drought policy pilot. Policy reform is a long-term task. However, in light of the clear need for adaptation policy to be cross-sectoral and thus for horizontal policy integration or 'mainstreaming' (Ahmad, 2009; Dovers and Hezri, 2010), the question arises of whether Australian policy addressed or was coordinated with connected impacts and policy sectors. In the space constraints here, we focus on one: Water allocation in the Murray-Darling Basin (MDB).

7.3 Drought and Water Allocation

The MDB is Australia's largest river basin and agricultural production area and depends on irrigation from regulated rivers. Post–World War II irrigation development (Smith, 1998) and weak and inconsistent policy settings across four state (provincial) jurisdictions allowed a situation of serious over-allocation to emerge, largely at the cost of environmental condition (Connell, 2007). Periodic water shortages from droughts led to a sequence of crises and policy responses in the early 1980s, 1990s and 2000s, with a history of a lessening of reform when climatic conditions ease. The creation of basin-wide intergovernmental arrangements in the 1980s was a pivotal achievement, as was the imposition of a cap on overall water use a decade later, deficiencies of these two measures notwithstanding. More fundamental from a public policy perspective was the inclusion of the water sector in profound micro-economic reforms in the mid-1990s under the umbrella of National Competition Policy via the Council of Australian Governments (CoAG). The 'CoAG water reforms' promoted greater self-reliance, proper pricing of water delivery, environmental water flows and water markets. As part of a strong process of economy-wide reforms, they were of longer-term impact than previous attempts.

Elements of this policy history were brought together in response to the unprecedented 2002 to 2008 drought. A series of intersecting intergovernmental agreements, monitoring and evaluation functions, information systems and institutional reforms have enabled and furthered the 2004 National Water Initiative (NWI), the world's most comprehensive water reform package (Hussey and Dovers, 2007; NWC, 2009; see www.nwc.gov.au). Water allocation under conditions of variability to achieve sustainable development balance across environmental, social and economic values is the core aim. Although national in a genuine sense, the overwhelming focus has been on the MDB as Australia's most troubled water system.

Key elements are:

- A comprehensive intergovernmental agreement and new national water laws, giving effect to statutory water planning, definition of sustainable limits to extraction and establishment of water markets as a key policy instrument
- Two independent statutory authorities, the National Water Commission (NWC) and Murray-Darling Basin Authority (MDBA)
- An unprecedented biennial assessment process (NWC, 2007; 2009) supported by significant advances in the quantity and quality of water and associated data
- A basin plan, to reform water allocations
- Significant funding aimed to encourage and enable reforms

Similar to drought policy, risk-sharing formulae have been established, based on the principle that water users should bear the risk of climate variability and future climate change, whereas governments are responsible for the impacts of policy change.

Redressing over-allocation and environmental degradation through limits to extraction, water planning, technological improvement, water markets and environmental flow provision must involve a reduction in overall use and re-allocation of water from consumptive to environmental uses. However, in 2010, the 'guidelines' for the Draft Basin Plan, incorporating significant allocation reductions in the order of those widely foreshadowed within the policy community, have been met with severe criticism that social and economic values had been overlooked in favour of environmental sustainability. The process of attempting to 'sell' the guidelines to affected agricultural producers in the Basin attracted a harsh response and a 'somewhat remarkable media frenzy' (Crase, 2011: 84). The public conversation failed to define over-allocation as a collective problem, and the matter was forced to be reviewed. Public meetings held throughout the Basin have served to dichotomise communities and the environment rather than link their futures. At the time of writing, this situation will (1) clearly take some time to resolve and may delay the MDB Plan, and (2) may even stall or reverse the policy directions built up over two decades. This situation might have been avoided had the policy process incorporated more recognition of the values of affected producers and communities.

Strangely, water reform does not deal directly with drought, even though climate variability is at the core, and drought policy does not deal with water allocation. The necessity of cross-sectoral (horizontal) policy integration has not been addressed. Similarly, industry policy over the 1980 and 1990s has been often unconnected to drought and water policy, as with structural adjustment measures that saw the dairy industry shift from (high-value) rain-fed coastal land to irrigated inland areas with resulting increased demands on MDB water. The imperative to horizontally coordinate policy is as intense as it is difficult in Australia as elsewhere (Ross and Dovers, 2008).

The MDB has been the major focus of water and therefore drought policy debates over recent decades, but the pattern is repeated elsewhere. The distinctly separate urban water policy domain has exhibited a similar history of reliance on 'predict-and-provide', supply-oriented approaches, punctuated by crisis management in drought and a lack of ongoing reform. In the wake of the 2002 to 2008 drought and the NWI, urban water has received much more attention, but also crisis-driven approaches dominated by the commissioning of large-scale desalination plants to augment supply, as well as some more adaptive management of behaviour and demand (Troy, 2008).

7.4 Lessons for Adaptation

In water and drought policy, as with disaster policy (Handmer and Dovers, 2013), a pattern of crisis-response-forgetting is common, as are the very real social and political pressures that inevitably (and properly) influence policy reform. If climate change results in more frequent and/or intense variability and thus drought, the difficulties for policymakers will intensify. So too will the likelihood of post-event regress. Drought in a variable climate is not a 'disaster' in the traditional sense, however disastrous the impacts on individuals, sectors and communities might be; the same is true for water shortages in a variable climate. Designing, implementing and especially maintaining policy regimes over the longer term is difficult.

Lessons from drought and water policy reform (and lack thereof) in the Murray-Darling Basin are relevant to other sectors. With increased variability, pressures to accept climate extremes and related disturbances to societies and environments, as expectable if not predictable, and to cater for these through robust, transparent policy settings will increase across events types. The alternative is *ad hoc* response to escalating impacts. The importance of understanding the policy history and path dependencies of specific sectors is important, as is the recognition of the complexities of public administrative, policy process, statutory and institutional mechanisms required. Ensuring cross-sectoral (horizontal) policy coordination, such as between drought and water, is a major challenge, as is cross-jurisdictional (vertical) coordination in a federal system such as Australia. The economic and political realities of resource- or place-dependent communities are highlighted, as is the importance of clear definitions of thresholds such as 'exceptional circumstances'. The use of triggers is a contentious policy issue, and there are arguments both for and against declarations as the basis for government intervention. The lessons from the National Drought Mitigation Center in the US suggest that the key to the implementation of workable, politically acceptable definitions is the development of triggers based on comprehensive, composite indicators that are developed in consultation with stakeholder groups.

However, there is a basis for believing this can be done, the difficulties noted here notwithstanding. Such policy development is consistent with policy trends in emergency management toward greater risk sharing and preparedness, with neoliberal-inspired public policy preferences toward self-reliance and risk management and with imperatives of policy integration. Much climate change adaptation can be addressed by 'normalising' adaptation, or increasing existing imperatives and capacities for reform across many sectors by incorporating the prospect of increased climate variability. The last challenge – cross-sectoral policy coordination – will be a particularly hard one. The future will be more of the same, but likely much more.

References

ACANZ – Agriculture Council of Australia and New Zealand (1992). *Record and Resolutions: 138th Meeting, Mackay 24 July 1992*. Canberra: Commonwealth of Australia.

Ahmad, I. H. (2009). *Climate Policy Integration: Towards Operationalization, Working Paper 73*. Washington, DC: United Nations Department of Social and Economic Affairs.

Botterill, L. C. (2003). Uncertain climate: the recent history of drought policy in Australia. *Australian Journal of Politics and History*, 49, 61–74.

Botterill, L. C. (2005). Late twentieth century approaches to living with uncertainty: the National Drought Policy. In *From Disaster Response to Risk Management: Australia's*

National Drought Policy, ed. L. C. Botterill and D. A. Wilhite. Dordrecht: Springer, pp. 51–64.

Botterill, L. C. (2006). Soap operas, cenotaphs and sacred cows: countrymindedness and rural policy debate in Australia. *Public Policy*, 1, 23–36.

Connell, D. (2007). *Water Politics in the Murray Darling Basin*. Sydney: Federation Press.

Courtice, B. (MP) (1989). *Appropriation Bill (No 3) 1988–89. Debate House of Representatives, 11 April 1989*. Canberra: Hansard.

Crase, L. (2011). The fallout to the guide to the proposed Basin Plan. *Australian Journal of Public Administration*, 70, 84–93.

Dovers, S. R. and Hezri, A. A. (2010). Institutions and policy processes: the means to the end of adaptation. *Wiley Interdisciplinary Reviews: Climate Change*, 1, 212–231.

Drought Pilot Review Panel (2011). *Review of the Pilot of Drought Reform Measures in Western Australia: Issues Paper*. Canberra: Commonwealth of Australia. Accessed 18 October 1012 from: <http://www.daff.gov.au/_data/assets/pdf_file/0011/1895852/wa-sub-cover-sheet.pdf>.

Drought Policy Review Expert Social Panel (2008). *It's About People: Changing Perspectives. A Report to Government by an Expert Social Panel on Dryness. Report to the Minister for Agriculture, Fisheries and Forestry*. Canberra: Commonwealth of Australia. Accessed 18 June 2013 from: <http://www.daff.gov.au/agriculture-food/drought/drought-program-reform/national_review_of_drought_policy/social_assessment/dryness-report>.

Drought Policy Review Task Force (1990). *National Drought Policy Volume I*. Canberra: Australian Government Publishing Service.

Handmer, J. and Dovers, S. (2013). *The Handbook of Disaster Policies and Institutions: improving emergency management and climate change adaptation*. 2nd Edition, London: Earthscan.

Hennessy, K., Fawcett, R., Kirono, D. et al. (2008). *An Assessment of the Impact of Climate Change on the Nature and Frequency of Exceptional Climatic Events*. Canberra: Commonwealth of Australia. Accessed 18 October 1012 from: <http://www.daff.gov.au/_data/assets/pdf_file/0007/721285/csiro-bom-report-future-droughts.pdf>.

Hussey, K. and Dovers, S. (eds.) (2007). *Managing Water for Australia: The Social and Institutional Challenges*. Melbourne: CSIRO Publishing.

NWC – National Water Commission (2007). *Australian Water Reform 2009: First Biennial Assessment of Progress in Implementation of the National Water Initiative*. Canberra: Commonwealth of Australia.

NWC – National Water Commission (2009). *Australian Water Reform 2009: Second Biennial Assessment of Progress in Implementation of the National Water Initiative*. Canberra: Commonwealth of Australia.

Productivity Commission (2009). *Government Drought Support: Productivity Commission Inquiry Report No 46, 27 February 2009*. Canberra: Commonwealth of Australia. Accessed 18 June 2013 from: <http://www.pc.gov.au/projects/inquiry/drought/report>.

Ross, A. and Dovers, S. (2008). Making the harder yards: environmental policy integration in Australia. *Australian Journal of Public Administration*, 67, 245–260.

Smith, D. I. (1998). *Water in Australia: Resources and Management*. Melbourne: Oxford University Press.

Troy, P. (ed.) (2008). *Troubled Waters: Confronting the Water Crisis in Australia's Cities*. Canberra: ANU e-Press.

8

After 'Black Saturday': Adapting to Bushfires in a Changing Climate

JOSHUA WHITTAKER, JOHN HANDMER,
AND DAVID JOHN KAROLY

On Saturday, 7 February 2009, 173 people lost their lives and 2,133 houses were destroyed by bushfires in the Australian State of Victoria (Figure 8.1; Teague et al., 2010). Fires burned under the most severe fire weather conditions experienced for more than one hundred years, with a record high maximum temperature of 46.4 °C in Melbourne, record low relative humidity and strong winds throughout the State (Karoly, 2009; National Climate Centre, 2009). The scale of life and property loss has raised fundamental questions about bushfire management and community safety in Victoria and throughout Australia. These include questions about Australia's 'prepare, stay and defend or leave early' policy, the adequacy of warning systems, the preparedness and responses of residents, fire authorities and other emergency services and the land-use planning system that manages development in high-fire-risk areas. These and other issues were investigated by the 2009 Victorian Bushfires Royal Commission, which handed down sixty-seven recommendations in its final report to the Victorian Government in July 2010 (Teague et al., 2010). Although the Commission heard evidence of the increased likelihood of extreme fire weather conditions because of climate change, none of its recommendations explicitly address climate change and its associated risks.

This chapter explores the environmental and human dimensions of the 'Black Saturday' bushfire disaster of 7 February 2009. It discusses factors influencing the severity of the fires, the vulnerability of people and property and changes to bushfire policy and management following the 2009 Victorian Bushfires Royal Commission. The chapter concludes with a brief reflection on the prospects for adapting to bushfires in a changing climate.

8.1 Environmental Factors

Bushfires are a regular occurrence in south-east Australia, associated with typical weather conditions including high maximum temperatures, strong winds, low relative

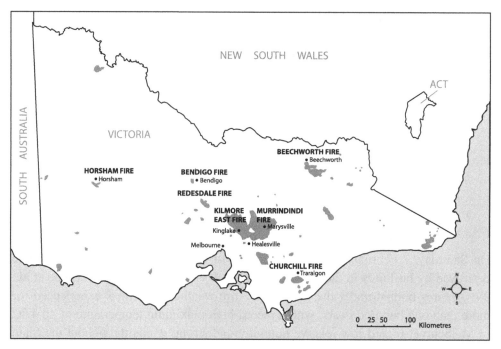

Figure 8.1 Location of the 2009 Victorian bushfires (January–February; shaded dark grey).

humidity and a preceding period of very low rainfall. Previous disastrous fires in south-east Australia occurred on Ash Wednesday, 16 February 1983, and Black Friday, 13 January 1939, both of which led to significant losses of life and property and changes to bushfire policy and management (see Pyne, 1998).

The McArthur Forest Fire Danger Index (FFDI) was developed in the 1960s as an empirical indicator of weather conditions associated with high and extreme fire danger and the difficulty of fire suppression (McArthur, 1967). The FFDI is the product of non-linear terms related to maximum temperature, relative humidity, wind speed and dryness of fuel (measured using a drought factor), each affecting the severity of bushfire conditions (Noble et al., 1980). The FFDI scale is used for the rating of fire danger and the declaration of total fire ban days in Victoria. It was developed so that the disastrous Black Friday fires in 1939 had an FFDI of 100.

Fuel load is not explicitly included in the FFDI, but it is crucial in determining the fire behaviour and intensity. Uncontrollable fire behaviour may occur at FFDI values of 100 for low fuel loads of only about 5 t/ha, much less than typical fuel loads of 12 t/ha for dry sclerophyll forest found in south-east Australia (Lucas et al., 2007). Fuel reduction burning prior to the Black Saturday fires in 2009 is believed to have inhibited the spread of several of these extreme fires (Teague et al., 2010).

To understand the environmental conditions associated with the catastrophic bushfires on 7 February 2009, we need to consider each of the factors in the FFDI.

Maximum Temperature

Melbourne and much of Victoria had record high maximum temperatures on 7 February 2009 (National Climate Centre, 2009). Melbourne set a new record maximum of 46.4 °C based on more than 100 years of observations. This new maximum was 0.8 °C hotter than the previous all-time record on Black Friday 1939 and 3 °C higher than the previous February record set on 8 February 1983 (the day of a dramatic dust storm in Melbourne). The urban heat island in Melbourne may have influenced these new records, but many other rural stations in Victoria set new all-time record maximum temperatures on 7 February. The high-quality rural site of Laverton, near Melbourne, set a new record maximum temperature of 47.5 °C, 2.5 °C higher than its previous record set in 1983. The extreme temperature on 7 February came after a record-setting heatwave ten days earlier, with Melbourne experiencing three days in a row with maximum temperatures higher than 43 °C during 28 to 30 January 2009, unprecedented in 154 years of Melbourne observations.

Drought Factor

Melbourne and much of Victoria had received record low rainfall at the start of 2009. Melbourne had thirty-five days with no measurable rain up to 7 February, the second longest period on record with no rain. The period up to 8 February, with a total of only 2.2 mm, was the driest start to the year for Melbourne in more than 150 years (National Climate Centre, 2009). This was preceded by twelve years of below-average rainfall over much of south-east Australia, with record low twelve-year rainfall over southern Victoria (National Climate Centre, 2009). This contributed to extremely low fuel moisture (3–5 per cent) on 7 February 2009.

Relative Humidity

Record low values of relative humidity were set in Melbourne and other sites in Victoria on 7 February with values as low as 5 per cent in the late afternoon. While long-term high-quality records of humidity are not available for Australia, the very low humidity is likely associated with the unprecedented low rainfall experienced from the start of 2009 in Melbourne and the protracted heatwave.

Wind Speed

Extreme fire danger events in south-east Australia are associated with very strong northerly winds bringing hot dry air from central Australia. The weather pattern and northerly winds on 7 February were similar to those on Ash Wednesday and Black Friday, and the very high winds do not appear to have been exceptional.

Table 8.1. *Forest Fire Danger Index (FFDI) and other environmental conditions for previous catastrophic fire events in Victoria. Temperature and rainfall were recorded in Melbourne*

Event	Date	FFDI	Maximum temperature (°C)	Rainfall for preceding 30 Days (mm)
Black Friday	13 Jan 1939	100	45.6	2.6
Ash Wednesday	16 Feb 1983	120	43.0	6.4
Black Saturday	7 Feb 2009	170	46.4	0

The combination of these factors led to unprecedented conditions for catastrophic fire danger on 7 February 2009 in Victoria, as shown in Table 8.1. Note that because of the non-linear and empirical relationships represented by the FFDI, this does not mean that the conditions for an FFDI of 170 are twice as bad as for an FFDI of 85, just substantially worse.

8.2 Human Factors

Victoria is recognised as one of the most fire-prone environments on earth because of its location, climate and vegetation (Luke and McArthur, 1978; Bradstock et al., 2002). However, human factors – such as ignition sources, the exposure of people and assets and fire and emergency management capabilities – are equally important in determining bushfire risk.

Like many major Australian cities, Melbourne has seen increasing residential development and population growth at its suburban fringes. According to the Australian Bureau of Statistics, the largest population growth for 2008 to 2009 occurred in five local government areas on Melbourne's outer suburban fringes (Australian Bureau of Statistics, 2010), with an already large population in the particularly high-risk areas to the north-east and east. Parts of regional Victoria have also experienced growth as an increasing number of people move away from metropolitan areas for the amenity and lifestyle of coastal and rural settings (Burnley and Murphy, 2004; Costello, 2007). Reduced housing affordability within inner Melbourne (Wood et al., 2008; Yates, 2008) has also encouraged residential development and population growth in high-bushfire-risk areas. These areas also contain critical water, electricity and telecommunications infrastructure that serves much of Melbourne.

With the extreme fire weather conditions discussed above, 7 February saw the ignition of more than 400 fires across Victoria. Most of the major fires were started by fallen power lines or arson (Teague et al., 2010). Fires quickly burned out of control as communities came under threat with little or no official warning. Firefighting capacities were limited in the face of such large and intense fires, leaving almost all residents to fend for themselves. Under the 'Prepare, stay and defend or leave early policy' (Australasian Fire and Emergency Service Authorities Council, 2005) adopted by all Australian fire services, residents were advised to prepare to stay and defend their home from bushfires or leave well before a fire arrived in their area. The policy was based on evidence that well-prepared people can safely protect well-prepared houses from bushfires and that a disproportionate number of deaths occur during late evacuations (see Handmer and Tibbits, 2005; Tibbits et al., 2008; Haynes et al., 2010). However, 173 people died in the 7 February fires – the largest number of fatalities in any Australian bushfire event – and more than 2,000 houses were destroyed. Almost immediately, the colloquially termed 'stay or go' policy was drawn into question (Australian Associated Press, 2009a), particularly considering police reports that 113 people died inside their homes (Australian Associated Press, 2009b). A number of studies provide insights into the human factors that contributed to the disaster.

Clearly, exposure to bushfire hazard was a major factor in the disaster. Crompton and colleagues (2010) found that 25 per cent of destroyed buildings in Marysville and Kinglake were located within the bushland boundary, while 60 per cent and 90 per cent were within 10 and 100 metres of bushland. The authors argue that distance to bushland was the most important factor influencing building damage in the fires. Nevertheless, interview and questionnaire research with households throughout the fire-affected areas, discussed below, found that many residents successfully defended their homes from the fires, including in Marysville and Kinglake.

A survey of 1,314 households affected by the fires (Whittaker et al., 2013) found varied levels of planning and preparedness, with residents in more suburban locations less likely to have considered themselves at risk from bushfires or to have taken measures to protect themselves. Half of those surveyed reported that they intended to stay and defend their homes and properties throughout the fire (50 per cent), with less than a fifth having a firm intention to leave *before* they came under threat (19 per cent). Significantly, more than a quarter of respondents intended to stay and defend but leave if they felt threatened (17 per cent) or wait and see what the fire was like before deciding whether to stay or leave (9 per cent). These 'wait-and-see' strategies are particularly risky and ill advised, as they increase the likelihood of late and dangerous evacuation.

Half of those surveyed actually stayed to defend their homes and properties from the fires (53 per cent). Of these, around one-third left at some stage during the fire because of perceived danger, the failure of equipment or utilities or because their house had caught fire. Of those who left before or when the fires arrived (43 per cent), more than half considered themselves to have left 'late' or 'very late' and more than three-quarters perceived the level of danger to be 'high' or 'very high' when leaving. Many reported experiencing difficulties associated with smoke, poor visibility, traffic, embers, flames and fallen trees. Most people did not receive an 'official' warning from police, fire or emergency services that a fire was threatening (62 per cent) but did receive 'unofficial' warnings from family, friends or neighbours (63 per cent).

Crucially, interviews revealed that even where warnings were provided and received, many waited until they were directly threatened before taking action.

Contrary to prevailing media coverage and popular opinion, many people stayed and successfully defended their homes and properties from the fires. The aforementioned survey of households from across the fire-affected areas found that rates of house damage and destruction were considerably lower among households where residents stayed and defended (Whittaker et al., 2013). In households where at least one person stayed to defend, just two in ten houses were destroyed. Half of the houses where all householders left or sheltered without defending were destroyed.

Handmer and colleagues (2010) reviewed each of the 172 civilian fatalities to identify factors that may have contributed to deaths. They found that around one-quarter of those who died had a chronic health condition that may have affected their mobility, judgement or stamina during the fire. Another 5 per cent were affected by an acute physical or mental condition that occurred on the day. Other factors found to have contributed to deaths included levels of planning and preparedness, with 58 per cent of those who died having made no apparent preparations and 53 per cent having no apparent fire plan or clear intention for what to do during a fire. The review found that more than two-thirds had been sheltering at the time they died (Handmer et al., 2010).

8.3 Changes to Bushfire Policy and Management

On 16 February, the 2009 Victorian Bushfires Royal Commission was established to investigate the circumstances leading to widespread losses of human life and property in the fires. Royal Commissions are major public inquiries established to advise government on important policy issues or to investigate allegations of impropriety, maladministration or major accidents (Prasser, 2006). The Commission's terms of reference were broad, covering a range of issues including the causes and consequences of the fires; the preparedness and responses of government, emergency services and households; measures taken to prevent disruption to essential services; and any other matters deemed relevant by the Commissioners (Teague et al., 2010). The Commission heard evidence from 434 experts and lay witnesses and received more than 1,200 public submissions. It handed down sixty-seven recommendations to government in its final report, including recommendations for bushfire safety policy, ignition prevention, emergency and incident management, land use planning and building, land and fuel management and organisational structure (see Teague et al., 2010 for a full list of recommendations). Although the Commission heard evidence of the increased likelihood of extreme fire weather conditions because of climate change, none of the recommendations explicitly address climate change and its associated risks. Here, we focus on recommendations we believe have the greatest potential to reduce losses of life and property in the future.

The Commission's first and most fundamental recommendation was for Victoria to revise its bushfire safety policy. Despite acknowledging that '. . . the central tenets of the stay or go policy remain sound', the Commission noted that the 7 February fires had exposed weaknesses in the way it was applied (Teague et al., 2010: 5). It found that the policy did not adequately account for 'ferocious' fires, which it said require a different kind of response. The Commission recommended a greater emphasis on the heightened risk to life and property 'on the worst days' and on leaving early as the safest option.

Given the unprecedented fire danger experienced on 7 February, the Commission's interim report (Teague et al., 2009) had recommended revision of fire danger ratings to include an additional rating beyond 'Extreme' and for existing ratings to be adjusted to correspond to higher Fire Danger Index values. In response, the National Framework for Scaled Advice and Warnings to the Community was developed (Australian Emergency Management Committee, 2009), which has seen the introduction of a 'Catastrophic' or 'Code Red' fire danger rating. In Victoria, the Country Fire Authority has developed scaled advice to more clearly communicate what residents should expect and do for different levels of fire danger (Table 8.2).

Other recommendations related to bushfire safety include the enhanced provision of timely and informative advice and warnings, improved advice on the 'defendability' of individual houses and an increase in the options available to people during fires, including community refuges and bushfire shelters (Teague et al., 2010). The latter recommendation reflected the Commission's view that the binary approach of 'stay and defend or leave early' does not adequately reflect the reality of what people do during fires: '. . . the reality [is] that people will continue to wait and see, and a comprehensive bushfire policy must accommodate this by providing for more options and different advice' (Teague et al., 2010: 5). In Victoria, the Country Fire Authority (CFA) and local governments have begun designating Neighbourhood Safer Places as places of last resort, while the Australian Building Codes Board is developing standards for the design and construction of bushfire bunkers for personal use. It is important to recognise, however, that although these changes expand the range of options available to people during bushfires, they do not eliminate the dangers associated with the failure to adequately plan and prepare, the tendency to wait and see and late evacuation. Nevertheless, the Commission recommended that authorities take greater responsibility for protecting those with limited capacities to protect themselves, for example, by developing plans for assisted evacuations (Teague et al., 2010).

The Royal Commission also offered a series of recommendations related to land use planning and building in high-bushfire-risk areas. The proposed changes aim to strengthen consideration of bushfire throughout the planning process in order to protect human life without imposing unacceptable biodiversity costs (Teague et al., 2010). They include strengthening the Wildfire Management Overlay – the main mechanism for controlling new developments and subdivisions in high-bushfire-risk areas – and a

Table 8.2. *The Country Fire Authority's scaled advice to the community (modified from Country Fire Authority, 2012)*

	What does it mean?	What should I do?
Code Red (FFDI 100+)	These are the worst conditions for a bush or grass fire. Homes are not designed or constructed to withstand fires in these conditions. The safest place to be is away from high-risk bushfire areas.	Leaving high-risk bushfire areas the night before or early in the day is your safest option – do not wait and see. Avoid forested areas, thick bush or long, dry grass. Know your trigger – make a decision about: when you will leave, where you will go, how you will get there, when you will return, and what you will do if you cannot leave.
Extreme (FFDI 75–99)	Expect extremely hot, dry and windy conditions. If a fire starts and takes hold, it will be uncontrollable, unpredictable and fast moving. Spot fires will start, move quickly and come from many directions. Homes that are situated and constructed or modified to withstand a bushfire, that are well prepared and actively defended, may provide safety. You must be physically and mentally prepared to defend in these conditions.	Consider staying with your property only if you are prepared to the highest level. This means your home needs to be situated and constructed or modified to withstand a bushfire, you are well prepared and you can actively defend your home if a fire starts. If you are not prepared to the highest level, leaving high-risk bushfire areas early in the day is your safest option. Be aware of local conditions and seek information by listening to ABC Local Radio, commercial radio stations or Sky News TV, go to cfa.vic.gov.au or call the Victorian Bushfire Information Line on 1 800 240 667.
Severe (FFDI 50–74)	Expect hot, dry and possibly windy conditions. If a fire starts and takes hold, it may be uncontrollable. Well-prepared homes that are actively defended can provide safety. You must be physically and mentally prepared to defend in these conditions.	Well-prepared homes that are actively defended can provide safety – check your bushfire survival plan. If you are not prepared, leaving bushfire-prone areas early in the day is your safest option. Be aware of local conditions and seek information by listening to ABC Local Radio, commercial radio stations or Sky News TV, go to cfa.vic.gov.au or call the Victorian Bushfire Information Line on 1 800 240 667.

Table 8.2 *(cont.)*

	What does it mean?	What should I do?
Very high/ High/Low-moderate (FFDI 0–49)	If a fire starts, it can most likely be controlled in these conditions and homes can provide safety. Be aware of how fires can start and minimise the risk. Controlled burning off may occur in these conditions if it is safe – check to see if permits apply.	Check your bushfire survival plan. Monitor conditions. Action may be needed. Leave if necessary.

'retreat and resettlement' strategy, including a scheme for non-compulsory acquisition of land by the state '. . . in areas of unacceptably high bushfire risk' (Teague et al., 2010: 33). The voluntary 'buy-back' scheme was the only recommendation rejected outright by the then-state government, on the grounds that vacant properties that were not maintained would increase fire risk for those who remained and that the cost would be prohibitive. The government also indicated that costs prohibited the replacement of all single-wire earth-return power lines with aerial bundled cable, underground cable or other technology to reduce the risk of bushfires (Willingham, 2010). A new state government, elected in November 2010, vowed to implement all of the Royal Commission's recommendations and has developed a voluntary 'bushfire buy-back' scheme to acquire high-risk properties from landholders who lost their primary place of residence in the fires (Department of Justice, 2012).

8.4 Adapting to Bushfires in a Changing Climate

Anthropogenic climate change is expected to continue to increase the frequency and severity of extreme fire danger in south-east Australia. In the Intergovernmental Panel on Climate Change (IPCC) Fourth Assessment Report of Working Group II, Hennessy et al. (2007) concluded that 'an increase in fire danger in Australia is likely to be associated with a reduced interval between fires, increased fire intensity, a decrease in fire extinguishments and faster fire spread' (515). A modelling study indicates that the number of Extreme fire danger days in south-east Australia is likely to increase by 15 to 65 per cent by 2020 relative to 1990 and by 100 to 300 per cent by 2050 for a high rate of global warming (Lucas et al., 2007; CSIRO, 2009).

In other fire-prone regions around the world, observed and expected increases in forest fire activity have been linked to climate change in the western US (Westerling et al., 2006), in Canada (Gillett et al., 2004) and in Spain (Pausas, 2004). While it is difficult to separate the influences of climate variability, climate change and changes in fire management strategies on the observed increases in fire weather, it is clear that climate change is increasing the likelihood of environmental conditions associated with extreme fire danger in south-east Australia (Lucas et al., 2007) and a number of other parts of the world.

As the frequency and severity of extreme fire danger increase, so too will the number of people and assets at risk. Victoria's population is projected to grow from 5.13 million in 2006 to 7.4 million in 2036 (Department of Planning and Community Development, 2009). Much of this growth will occur in coastal and inland areas near Melbourne and major regional centres (Department of Sustainability and Environment, 2005). The number of Victorian households is expected to grow even more rapidly, largely because of the ageing population, with an increase of 54.6 per cent from 2006 to 2036 (Department of Planning and Community Development, 2009). Consequently, there will be a greater concentration of people, homes, businesses and other assets in areas at risk from bushfire in the future.

The failure of land use planning to regulate development in high-bushfire-risk areas and the preparedness and responses of emergency services and residents have all been implicated in the Black Saturday bushfire disaster (Whittaker et al., 2009a, b; Buxton et al., 2010; Crompton et al., 2010). While the Royal Commission's recommendations and subsequent government and agency responses address these and many other issues, it is too soon to ascertain whether they will contribute meaningfully to adaptation. For example, it was envisaged that the introduction of Code Red warnings would alert residents to the potential for catastrophic bushfire and encourage them to leave early. However, research following a Code Red declaration during the 2009 to 2010 fire season found that very few residents actually left their homes (see Whittaker and Handmer, 2010). The increasing concentration of people and assets in areas at risk from bushfires and the difficulties of engaging and motivating people to protect themselves highlight the challenges facing the adaptation process.

References

Australasian Fire and Emergency Service Authorities Council (2005). *Position Paper on Bushfires and Community Safety*. East Melbourne: Australasian Fire and Emergency Service Authorities Council.

Australian Associated Press (2009a). 'Stay or go' policy to be reviewed. *The Age*, 9 February 2009. Accessed 22 August 2012 from: <http://www.theage.com.au/national/stay-or-go-policy-to-be-reviewed-brumby-20090209-81ek.html>.

Australian Associated Press (2009b). Black Saturday data reveals where victims died. *The Age*, 28 May 2009. Accessed 22 August 2012 from: <http://www.theage.com.au/national/black-saturday-data-reveals-where-victims-died-20090528-borp.html>.

Australian Bureau of Statistics (2010). *3218 – Regional Population Growth, Australia, 2008–09*. Canberra: Australian Bureau of Statistics. Accessed 12 August 2011 from: <http://www.abs.gov.au/ausstats/abs@.nsf/Products/3218. 0~2008–09~Main+Features~Victoria?OpenDocument>.

Australian Emergency Management Committee (2009). *National Framework for Scaled Advice and Warnings to the Community*. Canberra: Australian Government Attorney General's Department. Accessed 12 August 2012 from: <http://www.royalcommission. vic.gov.au/getdoc/891332ca-fc64-400c-b651-ee70f3d7ea15/RESP-7500-001-0016>.

Bradstock, R., Williams, J. E. and Gill, M. (eds.) (2002) *Flammable Australia: The Fire Regimes and Biodiversity of a Continent*. Cambridge, UK: Cambridge University Press.

Burnley, I. and Murphy, P. (2004). *Sea Change: Movement from Metropolitan to Arcadian Australia*. Sydney: UNSW Press.

Buxton, M., Haynes, R., Mercer, D. and Butt, A. (2010). Vulnerability to bushfire risk at Melbourne's urban fringe: the failure of regulatory land use planning. *Geographical Research*, 49, 1–12.

Costello, L. (2007). Going bush: the implications of urban-rural migration. *Geographical Research*, 45, 85–94.

Country Fire Authority (2012). *About Fire Danger Ratings*. Victoria: Country Fire Authority. Accessed 15 January 2013 from: <http://www.cfa.vic.gov.au/warnings-restrictions/about-fire-danger-ratings/>.

Crompton, R. P., McAneney, K. J., Chen, K., Pielke Jr, R. A. and Haynes, K. (2010). Influence of location, population and climate on building damage and fatalities due to Australian bushfire: 1925–2009. *Weather, Climate and Society*, 2, 300–10.

CSIRO – Commonwealth Scientific and Industrial Research Organisation (2009). *Climate Change and the 2009 Bushfires, CSIRO Submission 09/345 Prepared for the 2009 Victorian Bushfires Royal Commission*. Canberra: CSIRO Publishing. Accessed 12 August 2012 from: <http://www.royalcommission.vic.gov.au/Submissions/SubmissionDocuments/SUBM-002-031-0369_R.pdf>.

Department of Justice (2012). *Bushfire Buy-Back Scheme*. Melbourne: Department of Justice. Accessed 11 January 2012 from: <http://www.justice.vic.gov.au/buyback>.

Department of Planning and Community Development (2009). *Victoria in Future 2008: Victorian State Government Population and Household Projections 2006–2036*. Victoria: Department of Planning and Community Development. Accessed 12 August 2012 from: <http://www.dpcd.vic.gov.au/_data/assets/pdf_file/0008/32201/DPC056_VIF08_Bro_Rev_FA2.pdf>.

Department of Sustainability and Environment (2005). *Regional Matters: An Atlas of Regional Victoria*. Melbourne: State Government of Victoria.

Gillet, N. P., Weaver, N. J., Zwiers, F. W. and Flannigan, M. D. (2004). Detecting the effect of climate change on Canadian forest fires. *Geophysical Research Letters*, 31 (18), L18211.

Handmer, J. and Tibbits, A. (2005). Is staying at home the safest option during bushfires? Historical evidence for an Australian approach. *Environmental Hazards*, 6 (2), 81–91.

Handmer, J., O'Neill, S. and Killalea, D. (2010). *Review of Fatalities in the February 7, 2009, Bushfires*. Melbourne: Bushfire Cooperative Research Centre. Accessed 16 January 2012 from: <http://www.bushfirecrc.com/research/downloads/Review-of-fatalities-in-the-February-7.pdf>.

Haynes, K., Handmer, J., McAneney, J., Tibbits, A. and Coates, L. (2010). Australian bushfire fatalities 1900–2008: exploring trends in relation to the 'Prepare, stay and defend or leave early' policy. *Environmental Science and Policy*, 13, 185–94.

Hennessy, K., Fitzharris, B., Bates, B. C. et al. (2007). Australia and New Zealand. In *Climate Change 2007: Impacts, Adaptation and Vulnerability. Contribution of Working Group II to the Fourth Assessment Report of the Intergovernmental Panel on Climate Change*, eds. M. L. Parry, O. F. Canziani, J. P. Palutikof, P. J. van der Linden and C. E. Hanson. Cambridge, UK: Cambridge University Press, pp. 507–540.

Karoly, D. J. (2009). The recent bushfires and extreme heat wave in south-east Australia. *Bulletin of the Australian Meteorological and Oceanographic Society*, 22, 10–13.

Lucas, C., Hennessy, K., Mills, G. and Bathols, J. (2007). *Bushfire Weather in Southeast Australia: Recent Trends and Projected Climate Change Impacts. Consultancy Report Prepared for the Climate Institute of Australia*. Melbourne: Bushfire Cooperative Research Centre and CSIRO.

Luke, R. H. and McArthur, A. G. (1978). *Bushfires in Australia*. Canberra: Australian Government Publishing Service.

McArthur, A. G. (1967). *Fire Behaviour in Eucalypt Forest*. Canberra: Forestry and Timber Bureau.

National Climate Centre (2009). *The Exceptional January-February 2009 Heatwave in South-Eastern Australia*. Melbourne: Bureau of Meteorology.

Noble, I. R., Barry, G. A. V. and Gill, A. M. (1980). McArthur's fire-danger meters expressed as equations. *Australian Journal of Ecology*, 5, 201–203.

Pausas, J. G. (2004). Changes in fire and climate in the eastern Iberian Peninsula (Mediterranean Basin). *Climatic Change*, 63, 337–350.

Prasser, S. (2006). Royal Commissions in Australia: when should governments appoint them? *Australian Journal of Public Administration*, 65, 28–47.

Pyne, S. J. (1998). *Burning Bush: A Fire History of Australia*. Seattle: University of Washington Press.

Teague, B., McLeod, R. and Pascoe, S. (2010). *2009 Victorian Bushfires Royal Commission Final Report: Summary*. Victoria: State of Victoria.

Tibbits, A., Handmer, J., Haynes, K., Lowe, T. and Whittaker, J. (2008). Prepare, stay and defend or leave early: evidence for the Australian approach. In *Community Bushfire Safety*, eds. J. Handmer and K. Haynes. Collingwood: CSIRO Publishing, pp. 59–71.

Westerling, A. L., Hidalgo, H. G., Cayan, D. R. and Swetnam, T. W. (2006). Warming and earlier spring increase in western U.S. forest wildfire activity. *Science*, 313, 940–943.

Whittaker, J. and Handmer, J. (2010). Community bushfire safety: a review of post–Black Saturday research. *Australian Journal of Emergency Management*, 25, 7–13.

Whittaker, J., Haynes, K., Handmer, J. and McLennan, J. (2013). Community safety during the 2009 Australian 'Black Saturday' bushfires: an analysis of household preparedness and response. *International Journal of Wildland Fire*, http://dx.doi.org/10.1071/WF12010.

Whittaker, J., Haynes, K., McLennan, J., Handmer, J. and Towers, B. (2009a). *Research Results from the February 7th Fires: Second Report on Human Behaviour and Community Safety Issues*. Melbourne: Bushfire Cooperative Research Centre. Accessed 12 August 2012 from: <http://www.bushfirecrc.com/sites/default/files/managed/resource/mail-survey-report-10-1-10-rt-2.pdf>.

Whittaker, J., McLennan, J., Elliot, G. et al. (2009b). *Research Results from the February 7th Fires: First Report on Human Behaviour and Community Safety Issues*. Melbourne: Bushfire Cooperative Research Centre. Accessed 12 August 2012 from: <http://www.bushfirecrc.com/sites/default/files/managed/resource/chapter-2-human-behaviour.pdf>.

Willingham, R. (2010). Brumby rejects bushfire commissioner's land-buybacks, power-lines schemes. *The Age*, 27 August 2010. Accessed 22 August 2012 from: <http://www.theage.com.au/victoria/brumby-rejects-bushfire-commissioners-landbuybacks-powerlines-schemes-20100827-13urk.html>.

Wood, G., Berry, M., Taylor, E. and Nygaard, C. (2008). Community mix, affordable housing and metropolitan planning strategy in Melbourne. *Built Environment*, 34, 273–290.

Yates, J. (2008). Australia's housing affordability crisis. *The Australian Economic Review*, 41, 200–214.

9

Cyclone Tracy and the Road to Improving Wind-Resistant Design

MATTHEW MASON, KATHARINE HAYNES AND GEORGE WALKER

Early on Christmas morning 1974, tropical cyclone Tracy devastated the city of Darwin leaving only 6 per cent of the city's housing habitable and instigating the evacuation of 75 per cent of its population. The systematic failure of so much of Darwin's building stock led to a humanitarian disaster that proved the impetus for an upheaval of building regulatory and construction practices throughout Australia. Indeed, some of the most enduring legacies of Tracy have been the engineering and regulatory steps taken to ensure the extent of damage would not be repeated. This chapter explores these steps and highlights lessons that have led to a national building framework and practice at the fore of wind-resistant design internationally.

9.1 Cyclone Tracy

Tropical cyclone Tracy was a small but intense cyclone, with a landfall radius to maximum winds of 7 km, a forward speed of 7 km/h and central pressure of 950 hPa (Bureau of Meteorology, 1977) (Figure 9.1). Tracy was an Australian Category 4 cyclone with estimated maximum-gust wind speeds on the order of 250 km/h (70 m/s) (Walker, 1975). The recorded gust of 217 km/h (60 m/s) at Darwin Airport before the anemometer failed was, to that time, the highest wind speed measured anywhere on mainland Australia. Tracy's small size minimised the spatial extent of damage, but its slow translational speed meant areas impacted suffered more damage than might otherwise have been the case. Of cyclones that form in Australian waters, one passes within 200 km of Darwin every one to two years. The expected recurrence interval of an event similar to or stronger than Tracy impacting Darwin is greater than 100 years based on historical records.

Cyclone Tracy resulted in 71 fatalities and 650 injuries. Fortunately, flooding and storm surge were not major issues, or these numbers could have been far greater. The total insured loss from cyclone Tracy is reported at $200 million (1974 AUD), with an estimated total economic loss of around $500 to $600 million (1974 AUD) (Walker,

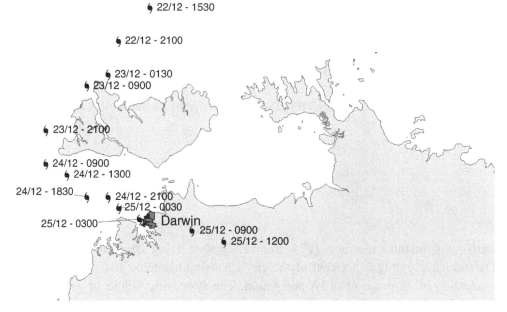

Figure 9.1 Tropical cyclone Tracy track from 22 to 25 December 1974.

2011). In addition, the immediate and prolonged evacuation of 35,000 residents (not to mention the event itself) led to significant and lasting psychological damage to many residents (e.g. Western and Milne, 1979).

9.2 Extent and Causes of Damage

Prior to cyclone Tracy, the structural design of housing in Darwin, as in the rest of Australia, was based on traditional forms of construction tempered by lessons learned from experience rather than rigorous engineering analysis. In contrast, all larger industrial and commercial (i.e. non-residential) buildings were specifically designed by structural engineers to resist cyclonic wind forces. The two contrasting approaches stemmed from the view that the cost of a single dwelling did not warrant the expenditure of ensuring structural adequacy against extreme events, while larger, more expensive structures did (Walker, 1975). There was, however, the belief, even among engineers, that methods used for housing construction in Darwin were adequate for cyclonic wind conditions, reflecting as they did lessons learned from cyclone Althea in Townsville three years earlier, even if they weren't specifically engineered as such. Cyclone Tracy showed this confidence to be misplaced.

Cyclone Tracy caused unprecedented damage to housing; 60 per cent of Darwin's houses were destroyed with only 6 per cent considered immediately habitable (e.g. Figure 9.2). Loss of roof cladding through fatiguing, the impact of flying debris and internal pressurisation were significant features of the observed damage (Walker,

Figure 9.2 Damage to houses in the northern suburbs of Darwin (photo by G. Walker).

1975). The fact that newer homes performed so poorly was a surprise to all given that their design incorporated engineering lessons from the damage to Townsville after cyclone Althea in 1971 (Walker, 2010). However, in accordance with the approach to housing at the time, the engineering was only applied to correct observed forms of failure, which, for example, did not include racking failures because of inadequate bracing, a mode of failure triggered in Darwin by the higher wind speeds. Furthermore, there had been a subtle change in the roof cladding material used after cyclone Althea; thinner high-strength steel, more prone to fatiguing failure under fluctuating wind loads, was substituted for the mild steel used previously. The consequences of this change had not been appreciated until Tracy showed it to be inadequate.

In contrast, larger engineered structures performed noticeably better, with only 5 to 10 per cent of buildings destroyed and around 80 per cent suffering minimal if any structural damage (Walker, 1975). The better performance occurred because adequate overall strength was built into the structure by engineers using legally enforced design codes. For those engineered buildings that did fail, construction errors or inappropriate application of design specifications were often to blame (n.b., some code inadequacies did exist).

Had housing performed better, the number of fatalities would have been reduced, and given that people would have had place to shelter after the storm, the size of the evacuation could have been reduced or even avoided altogether (Walker and Minor, 1980).

9.3 Adaptation of Building Design and Regulation

Cyclone Tracy was an engineering failure and required an engineering response. Answers to why buildings failed were needed, and any major reconstruction would have been misguided if failings were not acknowledged and accounted for. A moratorium was swiftly put on all rebuilding. Teams of engineers, architects and planners analysed most of the failed structures, determined failure causation and developed interim but conservative engineering-based design recommendations (the *Darwin Area Building Manual* published by the Darwin Reconstruction Commission in 1975) before reconstruction could begin. This process took less than three months, with the first design and construct tenders called in six and the first house completed within twelve. Taking a year to complete the first home was viewed by some as socially inadequate. The engineering input required to determine previously unknown failure mechanisms, develop and codify new testing and construction techniques, re-educate builders, inspectors and certifiers and then build the homes, however, warranted this time frame. A true social failure would have been reconstructing homes without addressing the reasons for the devastation that had just occurred.

Through the *Darwin Area Building Manual*, improved building standards were applied to large structures and housing alike. All homes rebuilt in Darwin were now legally required to abide by this manual and have a structural engineer responsible for their design. The relatively uniform design of most of Darwin's homes meant engineering costs were effectively spread over many homes. This was a huge step for the housing industry and signalled a paradigm shift in its direction. Housing was now considered at a comparable level of importance to larger buildings, and *all* buildings were now constructed with extreme events in mind.

Lessons learned during Tracy rapidly spread to other cyclone regions of Australia (Walker, 2010). The first implementation of engineering lessons outside Darwin was through the 1975 revision of the National Wind Loading Standard (AS1170.2, enforceable through state building by-laws), with further changes made in 1981. This standard initially only provided guidance for engineered buildings, but in 1977 a workshop incorporating members of the research community, government agencies and industrial organisations set out a series of guidelines so engineering lessons could be practically applied to the construction of housing. The resulting document, the Experimental Building Station's Technical Report (TR440, 1978), was widely used for the design and testing of residential construction in many of the country's cyclone regions. TR440 of itself was not a regulatory document; however, shortly after its introduction, a deemed-to-satisfy set of binding regulations for housing construction was set out for cyclone-prone parts of the State of Queensland in the Home Building Code, which was published as an appendix to the Queensland Government Building By-Laws in 1981. These regulations were again developed with input from a variety of sources, but at their heart were well-founded engineering principles. The Home Building Code was well received by the building industry, and its implementation replicated the paradigm shift of the *Darwin Area Building Manual* throughout Queensland.

The 1981 update to AS1170.2 and the contents of the Home Building Code were the last documents to have technical information directly attributable to cyclone Tracy (Reardon and Meecham, 1994). However, the changes instigated continued to evolve with nationally applicable deemed-to-satisfy housing design guidelines included in the 1989 revision of AS1170.2, which were then published as a stand-alone document, AS4055, in 1992 (updated in 2006) and called upon by all state building by-laws. The 1989 revision of AS1170.2 (updated in 2002 and 2011) also saw the implementation of a design philosophy (*limit state*) that accounts for extremes in a more scientific manner than the superseded process (*permissible stress*) (Walker, 2010). This change, though not directly instigated by, was the realisation of a major recommendation made in the aftermath of Tracy (Walker, 1975).

The manifestation of changes to building practice is the much-improved perform-ance of structures observed during more recent cyclones – for example, the absence of damage to homes built after the introduction of the Home Building Code in Kurrimine Beach during cyclone Winifred (1986). This is compared with significant damage to 20 to 30 per cent of the pre-code homes during the same storm (Reardon et al., 1986). Observations following cyclone Larry (2006) also indicate reductions of 20 to 60 per cent in the level of damage to newer homes (Henderson et al., 2006). Investigations following cyclone Yasi (2011) again found similar reductions (Boughton et al., 2011).

9.4 Lessons

Cyclone Tracy taught the engineering community many technical lessons. The dynamic nature of wind loading, impacts of flying debris, the role of internal pres-surisation and the importance of structurally engineering houses to resist wind were among the most valuable. Several other, more transferable lessons including the role of political and social factors in enabling or constraining adaptation were identified and are discussed below.

The Role of Scientific Input

The clearest lesson learned was that buildings designed using engineering principles performed considerably better than those that were not. The traditional trial-and-error or experience-based approach to housing construction was shown to be inadequate for regions subject to low-frequency, high-impact events in which little experience and/or feedback existed to help train the builder. The development and adoption of engineering-based building regulations for houses was an acknowledgement that for extreme events, practice and understanding had to be supplemented with additional information highlighting how severe a hazard could become and how the system

(i.e. the home in this case) would respond to this extreme load. The better performance of engineered buildings during Tracy and the subsequent good performance of
newer housing during recent cyclonic storms have shown that engineering experience
supplemented with up-to-date scientific understanding has led to greater structural
resilience.

The Role of Regulation and Building Codes

Building codes ensure a minimum level of design strength is maintained despite the
often limited exposure of an individual designer, builder or home owner to extreme
wind events. Building codes are ideal for turning knowledge into practice, and for the
case of low-frequency events, their regulated application is essential.

The *Darwin Area Building Manual* was effective in reducing the overall vulnerability in Darwin, but only because the city was largely rebuilt from scratch. For
rebuilding after a less intense cyclone in which the building stock sustains only minor
damage, building codes alone will not be an effective knowledge-transfer mechanism,
as their mandated use is only for new construction or substantial upgrades. A retrofit
program would be required in these cases to minimise community risk. The same
would apply for an attempt to reduce vulnerability in an established community.

In Australia, any retrofit program would probably need to be a regulatory requirement or be funded by government, as markets alone appear to be unable to incentiv-
ise people to act. In Fiji, however, incentives adopted by the insurance industry
more than twenty-five years ago have proved successful in improving building resilience without regulation and provide a counterpoint. Based on the Fiji experience,
detailed guidelines for retrofitting older houses in cyclone-prone areas in Australia
were developed by the insurance industry and published as HB 132 (HB132.1 Structural upgrading of older houses – Non-cyclone areas; HB132.2 Structural upgrading
of older houses – Cyclone areas, 1999) by Standards Australia, but as of yet have not
been fully utilised.

Opportunities and Conditions for Adaptation

Following most large disasters, political and economic imperatives mean that learning
is often not applied, is quickly forgotten or is poorly implemented. Public and political
pressure to rebuild and return life to 'normal' soon after a disaster means that delays
or changes are quickly overruled and rebuilding occurs in a similar vein to its pre-
disaster state. This effectively re-creates the conditions for disaster, as was the case
in the US after hurricane Camille in 1969 and the San Francisco earthquake of 1906
(Meyer, 2010). Fortunately, this wasn't the case in Darwin, where time was granted
for engineers to understand why failures occurred and to develop reconstruction procedures that overcame previous shortcomings. Very strong governmental commitment

was required for this to take place. Provided societal costs (psychological, social and economic) are accounted for and minimised, it seems prudent that this type of delay be repeated if appropriate remediation actions are not immediately clear.

The contextual circumstances within which building adaptations developed in Darwin were in many ways unique and perhaps simplified the implementation of a moratorium. Firstly, Darwin was an isolated city whose population had largely been evacuated, allowing the reconstruction delay with fewer objections than would likely be the case in other locations. In addition, federal control of much of the city's real estate and construction also reduced private and public pressures (Walker, 2011). It is unlikely that this context would exist for present-day disasters, and any reconstruction effort will face significant social and political hurdles that were avoided during Tracy. In light of this, retrofit or upgrade programs, such as discussed in the previous section, implemented prior to any disaster occurring are likely to be the most effective way to mitigate the impact of future disasters.

Inclusion and Education

Another factor in the successful transformation of the building industry was the involvement of researchers, designers, manufacturers, builders and inspectors in the process (Walker and Eaton, 1983). Involvement of people from all levels of the industry ensured that changes were soundly based and practically applicable. The high level of interaction also aided the dissemination of information and developed lasting relationships between parties, which aided continuing-education programs between levels.

9.5 Conclusions

Cyclone Tracy caused unprecedented damage to the city of Darwin and initiated a shift in the way housing is designed and built not only in Darwin but throughout Australia. The movement away from trial-and-error–based construction to a regulated, engineering-based codified design and construction method was a paradigm shift that significantly increased the resilience of Australia's housing stock.

The adaptation of building practice occurred only because there was strong public and political will, as is required for any regulatory change, for it to occur. Outside of a post-disaster time frame or in a location with strong local opposition to change, the extent and rapid implementation of such significant changes would have been impossible. Politically, allowing the right cross-section of people adequate time to develop appropriate solutions prior to any reconstruction was key to ensuring the long-term resilience of Darwin. The transfer of these lessons to other parts of the country through mandated implementation of building codes has minimised the risk that the devastation that beset Darwin will be repeated.

References

Boughton, G., Henderson, D., Ginger, J. et al. (2011). *Tropical Cyclone Yasi Structural Damage to Buildings, Cyclone Testing Station Technical Report 57*. Townsville: James Cook University.

Bureau of Meteorology (1977). *Report on Cyclone Tracy 1974, Technical Report 14*. Canberra: Australian Government Publishing Service.

Henderson, D., Ginger, J., Leitch, C., Boughton, G. and Falck, D. (2006). *Tropical Cyclone Larry: Damage to Buildings in the Innisfail Area, Cyclone Testing Station Technical Report 51*. Townsville: James Cook University.

Meyer, R. (2010). Why we still fail to learn from disasters. In *The Irrational Economist*, eds. E. Michel-Kerjan and P. Slovic. New York: Public Affairs Books, pp. 124–131.

Reardon, G. F. and Meecham, D. (1994). US Hurricanes of 1992: an Australian Perspective. In *Hurricanes of 1992*, eds. R. A. Cook and M. Soltani. New York: American Society of Civil Engineers, pp. 642–651.

Reardon, G. F., Walker, G. R. and Jancauskas, E. D. (1986). *The Effects of Cyclone Winifred on Buildings, Technical Report 27*. Townsville: James Cook Cyclone Structural Testing Station.

Walker, G. R. (1975). *Report on Cyclone "Tracy" – Effects on Buildings*. Canberra: Australia Department of Housing and Construction.

Walker, G. R. (2010). A review of the impact of Cyclone Tracy on building regulations and insurance. *Australian Meteorological and Oceanographic Journal*, 60, 199–206.

Walker, G. R. (2011). Comparison of the impacts of Cyclone Tracy and the Newcastle Earthquake on the Australian building and insurance industries. *Australian Journal of Structural Engineering*, 11 (3), 283–289.

Walker, G. R. and Eaton, K. J. (1983). Application of wind engineering to low rise housing. *Journal of Wind Engineering and Industrial Aerodynamics*, 14, 91–102.

Walker, G. R. and Minor, J. E. (1980). Cyclone Tracy in retrospect: a review of its impact on the Australian community. In *Proceedings, 5th International Conference on Wind Engineering*, ed. J. E. Cermak, vol. 2. New York: Pergamon Press, pp. 1327–1337.

Western, J. S. and Milne, G. G. (1979). *Social Effects of a Natural Hazard: Darwin Residents and Cyclone Tracy*. Canberra: Australian Academy of Science.

10

Adaptation and Resilience in Two Flood-Prone Queensland Communities

DAVID KING, ARMANDO APAN, DIANE KEOGH, AND MELANIE THOMAS

Floods are the world's second most costly natural hazard, averaging more than US$17 billion a year over the last decade, even though the death toll in most floods has not been as severe as their economic impact (CRED, 2009). Floods in Australia have cost, on average, AU$377 million per annum over recent decades (BITRE, 2008). In the summer and wet season of 2010 to 2011 in Queensland alone, floods were estimated to have resulted in AU$1.5 billion in insurance claims and an overall loss to the Australian economy reported to be at least AU$10 billion (ABC, 2011). During the 2011, floods thirty-five people died (Queensland Police, 2011) and 200,000 people were affected (BBC, 2010), including many thousands of people who may have evacuated in Queensland during the period of the floods and cyclones from late November to early February. Disastrous floods often prompt communities and governments to understand existing vulnerabilities and develop new policies and strategies to reduce vulnerabilities and prevent future disasters. The experience may offer insights to other communities as they experience future change.

We selected two Queensland communities that have experienced frequent flooding as case studies of adaptation and resilience (Apan et al., 2010). Both Charleville and Mackay have experienced regular riverine flooding in the past, although the floods considered here were unusual in that one was from a secondary tributary and one was flash flooding. Flash floods are common in Australia but present problems in terms of warnings because of the speed of their onset. The tragedy and loss of life from a flash flood in Toowoomba and the Lockyer Valley in south-east Queensland during 2011 demonstrated the risk of this hazard. The approach we undertook in the present case study was to reconstruct recent flood impacts in order to better understand how communities and businesses in these places have coped with disaster – their household and community resilience – as well as the measures that they had taken to mitigate against future floods. In addition, we consider their adaptive capacity in the face of longer-term change and flood impact. Although the case study research was carried out almost two years after the flood events, informants were able to reconstruct their experiences and reflect on their adaptation responses.

We used a questionnaire survey that was adapted from post-event instruments used by the Bureau of Meteorology, Geoscience Australia and the Centre for Disaster Studies. This enhanced comparison of two quite different settlements: A small inland outback town and a medium-sized coastal city. Three types of surveys were used: A structured random sample of households in flood-affected areas of each town, a random sample of business enterprises in the same regions and surveys and interviews of key informants drawn from government and non-government institutions in each town (Apan et al., 2010; Keogh et al., 2011).

10.1 The 2008 Floods in Charleville and Mackay

In the 2008 wet season, Queensland experienced heavy rainfall events that inundated approximately a million square kilometres of the state (62 per cent of its area) and cost the state and local governments approximately AU$234 million in damage to infrastructure. Many of the towns that were inundated in 2008 have experienced repeated flooding in subsequent years, culminating in the wet season of 2010 to 2011, when both riverine and flash floods occurred on an extensive scale throughout Queensland, causing severe inundation and disaster in many parts of the state. The 2008 case studies of Mackay and Charleville not only served as warnings of future floods in those towns but also anticipated the kind of disaster that unfolded extensively around Queensland in the 2010 to 2011 summer and wet season.

Charleville

The Warrego River, which flows alongside Charleville, has a well-documented history of flooding, with records of large floods dating back to 1910. More than ten major floods were recorded during this 100-year period, causing inundation of large areas that isolated Charleville from other towns and cities and caused major disruptions to road and rail links. The 2008 flood in Charleville, however, was the result of localised flooding along Bradley's Gully, which flows through the middle of the town, rather than flooding along the Warrego River. It was the biggest flood of Bradley's Gully since 1963.

Mackay

Similar to the Warrego River in Charleville, the Pioneer River that flows through the centre of Mackay has a history of flooding dating back to 1884. The highest flood, recorded in February 1958, peaked at 9.14 m. The February 2008 case-study flood, however, was a flash flood caused by intense local rainfall, with the river peaking at 7 m (Apan et al., 2010). On the morning of 15 February, more than 600 mm of rainfall was recorded in the lower Pioneer River catchment over approximately six hours. Intensity-frequency-duration analysis recorded by the Bureau of Meteorology of the rainfall in the area showed that the rainfall intensities significantly exceeded the 100-year average recurrence interval (GHD, 2009). Unofficial records of the total rainfall recorded over 24 hours for the Gooseponds Creek catchment (North Mackay) included 985 mm recorded at the suburb of Glenella (GHD, 2009), with a total of 886 residential properties inundated during the flood. Overall, in February 2008, the recorded rainfall for the city of Mackay was the largest on record.

10.2 Impacts of the 2008 Floods

Charleville

Flooding in Charleville during 1990 and 1997 was the impetus for the construction of Charleville's flood mitigation levee, which was almost completed prior to the 2008 flood. The levee has largely prevented flooding of the township from the Warrego River. However, during 2008, flood waters from Bradley Gully flowing through the township of Charleville flooded lower-lying properties.

Approximately forty businesses and residences in the lower-lying areas of Charleville were evacuated as well as some hospital patients. For safety reasons, power was cut to some areas (EMA, 2009). Following the 2008 Charleville flood, 920 families were assisted through Natural Disaster Relief and Recovery Arrangements (NDRRA) grants totalling more than AU$446,000 in Emergency Assistance and Essential Household Contents Grant payments (J. Peters, Community Recovery Unit, Queensland Department of Communities, Brisbane, personal communication, 23 December 2009). Concessional loans paid out to primary producers under NDRRA grants in Charleville of AU$658,000. Small business grants valued at AU$298,000 were also provided, while ninety-six primary producer grants valued at AU$1.341 million were paid out.

The total estimated cost of the January 2008 flooding in Charleville for the Department of Infrastructure and Planning for restoration of essential public assets for local government was AU$2.526 million; Emergency Management Queensland counter-disaster operations costs for Murweh Shire were AU$216,000, and restoration of essential public assets for state government was AU$482,000 (S. Hinkler, Queensland Department of Community Safety, personal communication, 18 January 2010).

Funding of AU$2.5 million was approved to reinstate the Murweh Shire road network under Natural Disaster Relief and Recovery funding (A. Pemberton, Murweh Shire Council, personal communication, 2 November 2009). The hospital's emergency department recorded a 22 per cent increase in numbers of patients in the March 2008 quarter.

Mackay

In Mackay, flood waters impacted approximately 4,000 homes. Schools were shut, the local road network was badly damaged, more than 6,200 homes lost power and mobile and land line communications were disrupted. One person died (a 17-year-old man) when he disappeared in the Pioneer River. Mackay Airport was closed and SES crews answered 2,000 calls for assistance. Six evacuation centres were established. For NDRRA grants, a total of 5,369 Emergency Assistance Grants (AU$1,996,450) and 1,512 Essential Household Contents Grant applications (AU$2,334,002) were provided. More than 5,400 families were assisted in the Mackay region (J. Peters, Community Recovery Unit, Queensland Department of Communities, Brisbane, personal communication, 24 December 2009).

Small business and primary producer grants came to AU$9.9 million and estimated costs for the Department of Infrastructure and Planning, Road Base Saturation and Emergency Management Queensland counter-disaster operations and restoration of essential public assets AU$39.145 million. The total cost of general insurance claims paid out was approximately AU$410 million. These claims related to items such as damage to building and contents, motor vehicles, business interruption, fencing in rural areas and the like. Impacts such as these are direct measures of the vulnerability of the towns to the flood hazard.

10.3 Recognising Vulnerability

The physical vulnerability of both Charleville and Mackay is a consequence of their proximity to highly flood-prone rivers in catchments that are also subject to rapid inundation and at risk of flash flooding. Levees have been constructed in both towns to restrict river floods, but both of the 2008 events occurred inside the levees, where barriers limited the capacity of flood water to escape into the main river. This physical vulnerability remains in each locality.

Social and community vulnerability is often driven by issues of equity. Societal vulnerability is a consequence of structural factors of the economy, social class and ethnicity, frequently exacerbated by personal factors such as marital and family status (Blaikie et al., 1994; Anderson-Berry and King, 2005). A consequence of social inequality is the development of areas of towns that are socially disadvantaged (Wisner et al., 2011). Low rental areas that are occupied by disadvantaged groups, including the poor and the elderly, are often in flood-prone parts of the town, as is the case in both Mackay and Charleville. Social vulnerability factors are unavoidable in the short term and are slow to change.

The flood event in Mackay resulted in mass evacuation with admission to hospital of people requiring special care. In Charleville, there was a lack of suitable accommodation for nursing home evacuees. Additionally, Charleville is a remote location, which constrains transport to other towns in the region and restricts communication.

The household surveys indicated that in both Mackay and Charleville, there were relatively low levels of preparation for the flood, primarily because these were rapid-onset flash-flood events. While both towns are clearly vulnerable to flooding and its impacts, they also demonstrated strong levels of resilience and capacity to mitigate the impacts of floods (see next section).

10.4 Building Resilience and Mitigation

Mitigation – the reduction of the physical flood threat – occurs at multiple levels, including government and council, businesses and the household level. Both Charleville's and Mackay's planning departments had flood overlays and hazard land use zones at the time of the floods that provided a framework to constrain future urban growth and inform residents of risk and trigger evacuation warnings (following the 2011 floods, additional mapping is underway). However, state planning policy emphasises river flooding rather than flash floods that are independent of river levels, and there is currently no mechanism for assessing the extent to which developments contribute towards the infilling of floodplains and therefore contribute to increased flash flooding (Thomas et al., 2011). This sort of planning and urban development issue emerged from the 2011 Queensland Flood Inquiry (QFCI, 2011) and was addressed in its final report (QFCI, 2012).

Mitigation measures undertaken by households and businesses enhanced community and household resilience during the 2008 floods. In particular, insurance coverage assisted the recovery process, although businesses were generally better covered than households. Practical measures such as the maintenance of property drains and ditches and the placing of valuable items above ground level were successfully practised by both households and businesses, although for retail establishments, the need to be at street level for customer access posed constraints in preventing drains immediately in front of premises facilitating ingress of flood waters.

Household surveys identified existing social capital (positive and supportive social and civic relationships as defined by Putnam (2000)) and social networks (Moore et al., 2003; Barney, 2004) in both communities, especially organisational memberships that include activities such as volunteering (Hall and Innes, 2008). Both social capital and networks are indicators of resilience. These indicators were identifiably more active in Charleville, which possessed many community organisations that reinforced strong social bonds and networks. Mackay, on the other hand, had a greater range of opportunities but much looser networks that gave an impression of lower levels of resilience than Charleville, but this arose partly because of the research problem of identifying fewer tangible social groups and connections. For example, following the Mackay floods, the evacuation of households was for extended periods, in some cases up to six months before returning home, yet those households that were evacuated made very little use of institutional evacuation centres. Most people stayed with friends or relatives or rented short-term accommodation – actions that are indicators of community resilience.

10.5 Strategies for Adaptive Capacity and Community Viability

If floods continue to have impact upon a community and result in a regular pattern of loss and recovery, resilience and hazard mitigation may not be sufficient protection. Relocation elsewhere is a rational adaptive strategy. Migration has been identified as a climate change adaptive strategy operating at individual and household levels (Handmer et al., 1999; IPCC, 2007). The household and business surveys carried out in Mackay and Charleville asked people how probable it might be that they would consider moving either to another part of the town or to a different town in the event of repeated severe flood events (Table 10.1).

Although the proportions of households or businesses that might relocate are relatively small, the compounding impact of such out-migration could lead to long-term decline in population. In Table 10.2, we cross-tabulate the percentage of people who responded that in the event of more flood events in the future, they might consider relocation within the town or migration to another town, according to socio-economic characteristics that were recorded in the household survey. Of those who had been less than 5 years in the community, 11 per cent considered moving elsewhere in town; of those who had been resident between 6 and 10 years, 25 per cent considered moving within the town; and for residents of more than 10 years, 29 per cent considered relocation as a strategy. The length of residency, which may be considered as a measure of resilience, is negatively related to adaptation by internal relocation.

Few householders considered relocating to a different town; 17 per cent of new residents (less than 5 years), 11 per cent of medium-term residents (6–10 years) and

Table 10.1. *Likelihood that in the event of further floods, businesses and households would consider the potential for relocation identified from household and business surveys (Apan et al., 2010; Keogh et al., 2011)*

	Intentions of businesses		Intentions of households	
	Mackay %	Charleville %	Mackay %	Charleville %
Move to a different part of town				
Not likely to be considered	64.3	100.0	52.0	57.4
Neutral	14.3	–	25.3	11.1
Likely to be considered	21.4	–	22.6	31.5
Move to a different town				
Not likely to be considered	92.7	81.8	69.4	77.8
Neutral	7.3	–	16.0	9.3
Likely to be considered	–	18.2	14.7	13.0

Table 10.2. *Socio-economic indicators that influence relocation as an adaptation response from household surveys (Apan et al., 2010; Keogh et al., 2011)*

Socio-economic and demographic characteristics	Response in event of more frequent floods	
	Consider relocation within the town %	Consider leaving the town %
Resident < 5 years	11	17
Resident 6–10 years	25	11
Resident > 10 years	29	15
Family with children	29	12
Couple	27	14
Living alone	12	18
Indigenous	33	17
Non-Indigenous	25	13
Employed full time	31	15
Unemployed	17	14
No formal qualifications	31	15
Trade qualification	29	13
Tertiary qualification	23	17
Overall response Mackay	**23**	**15**
Overall response Charleville	**32**	**13**

15 per cent of long-term residents (more than 10 years). Hence, relocating away from flood-prone locations or suburbs is a much stronger adaptation strategy than leaving the town altogether, with willingness to relocate increasing with length of residency (Table 10.2).

The primary family types shown in Table 10.2 illustrate opposite trends for relocation either within the town or departure from the town. (The category of 'living alone' includes many elderly who were probably immobile as well as young people who are more mobile.) It is interesting that indigenous respondents showed a greater propensity to relocate than non-indigenous residents. This may relate to extended family networks in other settlements. For example, after cyclone Larry devastated the Innisfail area in 2006, many indigenous banana-farm workers left the region to return to 'home' communities (Glick, 2006).

In the case of Charleville, relocation within the town is limited because of its small size, especially for businesses that are almost entirely located in the town centre. In both towns, a small proportion of both businesses and households suggested that they would seriously consider relocating to another town if further severe floods reoccurred. Although the proportion is a minority of the population, such an emigration outflow would have a significant effect on the economy, growth rate and long-term viability of either town. Anecdotal comments from survey respondents suggested that significant numbers of people had already left the town, especially Mackay, after the

2008 floods, but there is a problem of ascribing this to the impact of the flood alone because of the economic downturn of the global recession at the end of that year. Following the Queensland floods of 2010 to 2011, similar reports of out-migration are widespread, and again it is mixed up with the consequent economic impact of the disaster.

Respondents in both towns suggested that they had increased preparations for future floods and that they might be better capable of adapting to future disasters. While there was both a willingness and capacity to seek information about floods and preparedness, there was a clear tendency on the part of both households and businesses to consider flood preparation to be the primary responsibility of government and in particular local government. This highlights the notion of institutional dependency, a factor that may work against adaptation. Another barrier to adaptation cited by both household and business respondents was the high costs associated with flood mitigation. People reported that they had taken out more flood insurance but that they had initiated very few structural changes to buildings or property.

Both the population and the economy of Mackay are more diverse than in Charleville (Mackay, with 118,842 people and a gross regional product of AU$5.1 billion, contrasts with Charleville's Murweh Shire population of just 4,910 and an agriculture worth AU$60 million, comprising 50 per cent of the economy). A larger centre is economically more resilient, with a greater adaptive capacity than a small community, simply because of the greater choice and the larger numbers of people, businesses and infrastructure as well as more access to government resources.

Individuals in small communities and the communities themselves may possess stronger social capital than people and communities within larger cities, but the diversity and quantity of opportunities in cities give them an overall advantage in terms of resilience and adaptive capacity. West has shown that as they expand, cities multiply their opportunities at a greater rate than that of their population growth. 'With each increase in size, cities get a value-added of 15%. Agglomerating people, evidently, increases their efficiency and productivity Cities create problems as they grow, but they create solutions to those problems even faster, so their growth and potential lifespan is in theory unbounded' (West, 2011; also see Bettencourt et al., 2007).

10.6 Conclusion: Policy, Legislation and Community Adaptation Strategies

Charleville may be closer to the edge of population viability than the more diverse and robust but no less flood-vulnerable coastal city of Mackay, because it is a small remote town in a region that has been negatively impacted by drought and structural economic changes. Although the research showed Charleville possessed stronger social capital than Mackay, demonstrating characteristics of resilience that are likely to contribute to its long-term adaptive capacity, without the nurturing of the social capital strengths of each community, there is a probability that repeated flood disasters will undermine their viability, particularly if businesses and households relocate. In that case, the greater diversity and choices of the city will very likely overshadow the strong social capital of the small rural town, giving the city communities the edge in resilience and adaptive capacity simply because the city is so much greater than the sum of its individuals' resilience.

Adaptive capacity is nurtured through social capital, strong community networks, community organisations and volunteering and probably most of all through a diversity of people and social groups. Specific policy interventions that may contribute to the enhancement of social capital to build adaptive capacity include support for community organisations and volunteering, special-needs registers, strengthening of the state planning policy in order to empower planners in making appropriate land-use decisions and the requirement that rebuilding of flood-damaged structures should be to flood zone standards that include building codes as well as restrictions on redevelopment in the most flood-prone locations. Some of these solutions are now being discussed following the 2010 and 2011 floods.

Emergency management also needs to adapt to new types of local hazards such as more frequent severe flash floods and will need to innovate in the area of its own personnel flexibility and training in order to adapt to increased impacts, events that occur with limited warnings and social changes that may accompany climate change in the future.

A significant social change that has been identified in this case study is the probability of relocation and migration. Within larger settlements and cities, much relocation may take place within existing urban boundaries, but for smaller towns, where there are limited options, increased flood hazards that accompany climate change may contribute to long-term and absolute population loss. Policymakers need to understand and model the implications of movements of people as an adaptation response to climate change in order to plan for both community viability and a significant alteration in population distribution.

References

ABC – Australian Broadcasting Corporation (2011). Flood costs topped to top $30b. *ABC News*, 18 January 2008. Accessed 16 May 2012 from: <www.abc.net.au/news/stories/2011/01/18/3115815.htm>.

Anderson-Berry, L. and King, D. (2005). Mitigation of the impact of tropical cyclones in Northern Australia through community capacity enhancement. *Mitigation and Adaptation Strategies for Global Change*, 10, 367–392.

Apan, A., Keogh, D. U., King, D. et al. (2010). *The 2008 Floods in Queensland: A Case Study of Vulnerability, Resilience and Adaptive Capacity, Final Report*. Gold Coast, Australia: National Climate Change Adaptation Research Facility.

Barney, D. (2004). *The Network Society*. Cambridge, UK: Polity Press Ltd.

BBC – British Broadcasting Corporation (2010). Australia: Queensland floods spur more evacuations. *BBC News*. 31 December 2010. Accessed 16 May 2012 from: <www.bbc .co.uk/news/world-asia-pacific-12097280>.

Bettencourt, L. M. A., Lobo, J., Helbing, D., Kuhnert, C. and West, G. B. (2007). Growth, innovation, scaling, and the pace of life in cities. *Proceedings of the National Academy of Sciences of the United States of America*, 104(17), 7301–7306.

BITRE – Bureau of Infrastructure, Transport and Regional Economics (2008). *About Australia's Region: June 2008*. Canberra: Australian Department of Infrastructure, Transport Regional Development and Local Government.

Blaikie, P., Cannon, T., Davis, I. and Wisner, B. (1994). *At Risk: Natural Hazards, People's Vulnerability and Disasters*. London: Routledge.

CRED – Centre for Research on the Epidemiology of Disasters (2009). *Emergency Events Database (EM-DAT)*. Brussels: Centre for Research on the Epidemiology of Disasters. Accessed 1 December 2012 from: <http://www.emdat.be/>.

EMA – Emergency Management Australia (2009). *Disasters Database – Widespread Flooding: Queensland*. Canberra: Emergency Management Australia. Accessed 20 December 2012 from: <http://www.em.gov.au/Resources/Pages/DisastersDatabase .aspx>.

GHD (2009). *Gooseponds and Vines Creek Flood Study: Final Report*. Mackay: Mackay Regional Council.

Glick, J. D. (2006). *Banana Farms in the Innisfail Region Eight Months after Cyclone Larry: Dynamics of Farm-Level Recovery; December 2006*. Townsville: Centre for Disaster Studies, James Cook University.

Hall, J. and Innes, P. (2008). The motivation of volunteers: Australian surf lifesavers. *Australian Journal on Volunteering*, 13(1), 17–21.

Handmer, J., Dovers, S. and Downing T. (1999). Societal vulnerability to climate change and variability. *Mitigation and Adaptation Strategies for Global Change*, 4, 267–281.

IPCC – Intergovernmental Panel on Climate Change (2007). *Climate Change 2007: Synthesis Report. Contribution of Working Groups I, II and III to the Fourth Assessment Report of the Intergovernmental Panel on Climate Change*, eds. Core Writing Team, R. K. Pachauri and A. Reisinger. Cambridge, UK: Cambridge University Press.

Keogh, D. U., Apan, A., Mushtaq, S., King, D. and Thomas, M. (2011). Resilience, vulnerability and adaptive capacity of an inland rural town prone to flooding: a climate change adaptation case study of Charleville, Queensland, Australia. *Natural Hazards*, 59, 699–723.

Moore, S., Eng, E. and Daniel, M. (2003). International NGOs and the role of network centrality in humanitarian aid operations: a case study of coordination during the 2000 Mozambique floods. *Disasters*, 27(4), 305–318.

Putnam, R. (2000). *Bowling Alone: The Collapse and Revival of American Community*. New York: Simon and Schuster.

QFCI – Queensland Floods Commission of Inquiry (2011). *Queensland Floods Commission of Inquiry Interim Report*. Brisbane: Queensland Floods Commission of Inquiry.

QFCI – Queensland Floods Commission of Inquiry (2012). *Queensland Floods Commission of Inquiry Final Report*. Brisbane: Queensland Floods Commission of Inquiry.

Queensland Police (2011). *Death toll from Queensland floods*, media release, Queensland Police, Brisbane, 24 January. Accessed 16 May 2012 from: <http://www.police.qld.gov. au/News+and+Alerts/Media+Releases/2011/01/death_toll_jan24.htm>.

Thomas, M., King, D., Keogh, D. U., Apan, A. and Mushtaq, S. (2011). Resilience to climate change impacts: a review of flood mitigation policy in Queensland, Australia. *Australian Journal of Emergency Management*, 26(1), 8–17.

West G. (2011). *Why Cities Keep on Growing, Corporations Always Die, and Life Gets Faster*. Presentation at Cowell Theatre in Fort Mason Center, San Francisco, California, 25 July 2011. Accessed 12 December 2012 from: <http://longnow.org/seminars/02011/jul/25/why-cities-keep-growing-corporations-always-die-and-life-gets-faster/>.

Wisner, B., Gaillard, J. C. and Kelman, I. (eds.) (2011). *Handbook of Hazards and Disaster Risk Reduction*. London, UK: Routledge.

Part III

Case Studies from Europe

11

Windstorms, the Most Costly Natural Hazard in Europe

UWE ULBRICH, GREGOR C. LECKEBUSCH, AND MARKUS G. DONAT

11.1 The Role of Windstorms in Europe

Windstorms cause more than half of the economic loss associated with natural disasters in Europe (Munich Re, 1999; 2007). Their comparatively high frequency in combination with the enormous concentration of values in Europe results in a loss potential only comparable with earthquakes and hurricanes in the United States and Japan. Consequently, there is a high awareness of these risks in European countries. The institutions involved in the warning and rescue processes, like weather services, fire brigades and regional governments, have developed methodologies along the applicable regulations, which are country specific. Risk map zones have been developed and have been incorporated in official building regulations (e.g. DIN 1055–4 in Germany). Insurance against the monetary loss arising from the direct effects of windstorms is not compulsory in all countries. Still, such insurance is commonly contracted. With respect to residential buildings, for example, there is a typical insurance density of about 90 per cent (Bresch et al., 2000). The cumulative effects of storm damage could endanger insurance company solvency, which is again subject to national and international regulations (Solvency II). Part of the required knowledge of windstorm risks is based on proprietary risk models, available to insurance brokers on a commercial basis from companies specialised in these services.

11.2 Identification of Windstorms

Large-scale windstorm events can be identified either from the damage they produce or from meteorological data. Of course, the representation of a windstorm in terms of meteorological data and damage data may be quite different. A top storm in meteorological terms is not always a top storm in terms of damage (and vice versa), as significant damage sums only occur when high winds hit a highly populated area. Such effects must be taken into account, in particular when attempting to use meteorological data for assessments of related damage.

Major past storm events in terms of losses have been included in lists and statistical evaluations of individual insurance companies (in particular of the major Re-insurance and broker companies such as Munich Re, Swiss Re, Partner Re, etc.), which are partly made available for the public and for public research (cf. Bresch et al., 2000). Insurance companies have also developed a definition of a severe storm. In order to prevent fraud, a rule is required defining how strong winds have to be before customers are entitled to apply for compensation. In Germany, for example, the wind speed threshold is set to Beaufort (Bft) wind force number 8 (sustained wind speeds > 17.2 m/s).

A meteorological identification of windstorms requires a definition distinguishing an extreme event from 'normal' intense near-surface winds. Typical wind speeds near the surface (usually measured at 10 m above the ground) are strongly dependent on both the surface characteristics and boundary layer conditions (Goyette et al., 2003; Hofherr and Kunz, 2010). The highest wind speeds associated with windstorms generally occur over water, while winds being dragged over rougher land surfaces tend to be slower. Also, climate models usually do not produce wind speeds exceeding Bft 8 (Rockel and Woth, 2007), which makes relative criteria necessary for the identification of windstorms in this kind of data. To assess the storm intensity relative to the local wind climatology, the systematic effect of roughness should be eliminated in order to determine a storm field. This can be achieved applying quantile-based approaches (e.g. Della-Marta et al., 2009), that is, by identifying the areas in which wind speeds exceed a certain local (surface-roughness-specific) percentile. Following Klawa and Ulbrich (2003), several researchers have used the 98th percentile of daily maximum wind speeds as the threshold for a definition of extreme wind events (e.g. Schwierz et al., 2010). For Germany, this choice is motivated by insurance practice, as Bft force 8 is roughly equivalent to the 98th percentile of wind speed over the plain areas in northern Germany. An example of such a wind field is given in Figure 11.1. Subsequent studies have confirmed that this choice is suitable for estimations of storm damage, as loss computed from a storm loss model based on this threshold is able to reproduce insurance loss data using either station wind-speed data (Klawa and Ulbrich, 2003) or reanalysis data (e.g. Pinto et al., 2007b; Donat et al., 2010a) (Figure 11.2). Other studies have identified windstorms using percentiles of sea level pressure (SLP) gradients between stations as a basis (Alexandersson et al., 2000; Wang et al., 2009), as measurements of SLP are generally considered less influenced by inhomogeneities than wind measurements.

A typical large-scale European storm should be distinguished from local wind extremes occurring, for example, in conjunction with thunderstorms. The large areas typically affected by windstorms contribute to the high-cumulated loss amounts caused by those events. Consequently, Leckebusch and colleagues (2008a) identify windstorms from a minimum area simultaneously hit by extreme winds (defined by exceedance of the local percentile) and a minimum duration of 24 hours, as these choices led to the identification of the main events known from insurance, excluding small-scale events from storm climatologies.

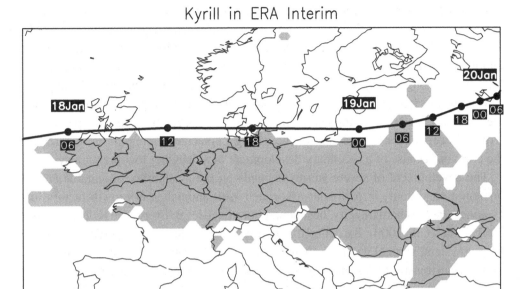

Figure 11.1 Storm Kyrill causing extreme wind speeds over large parts of Europe in January 2007. The track of the cyclone (bold line) and the areas where extreme wind speeds occur (exceeding the local 98th percentile, grey shadings) are shown. The cyclone track and extreme wind speeds were calculated from ERA-Interim reanalysis.

11.3 Physical Processes Leading to Windstorms

Large-scale European winter storms are a phenomenon related to the Atlantic storm track, which is one of the maxima of synoptic-scale variability in the Northern Hemisphere (e.g. Christoph et al., 1995). These winter storms are produced by travelling low- and high-pressure systems at the surface and associated troughs and ridges in the upper troposphere. Europe finds itself geographically downstream of the maximum

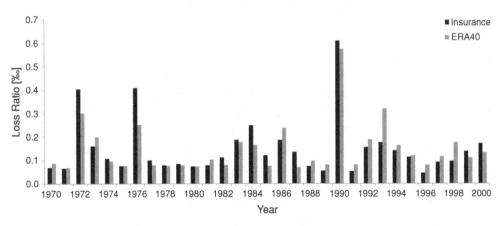

Figure 11.2 Annual loss ratio for Germany (‰) as recorded by the German Insurance Association and calculated by applying a storm loss model to ERA40 reanalysis wind fields (modified from Donat et al., 2010a).

of the Atlantic storm track located over Newfoundland. The surface cyclones affecting northern and central Europe typically have their genesis region partly over North America and the North Atlantic, in particular in the vicinity of Greenland, while a major part of the systems affecting southern Europe is produced in the area itself. Extremely strong surface winds associated with the cyclones occur when local pressure gradients (mostly south of the cyclone core) become large. The storm events are usually related to a significant intensification of the cyclones over Europe (e.g. in terms of deepening core pressure or increasing Laplacian of pressure in the area of the core) or because of a secondary development of a daughter low-pressure system.

The development of severe storms depends on the availability of suitable environmental conditions, in particular associated with anomalies in growth factors for cyclones, such as baroclinicity, heat content or upper-level divergence (Ulbrich et al., 2001; Pinto et al., 2009). Baroclinicity quantifies the meridional temperature gradient (or, equivalent, vertical wind shear) in the mid-latitude troposphere, which is reduced by the heat transports associated with cyclones. Heat content anomalies, combined from sensible and latent heat contributions, are a factor for upward winds, which in turn can enhance cyclone intensity and hence local winds at the surface, as also upper-level divergence (for example, in the exit regions of upper-level jets). Ageostrophic wind components arising from the movement of the pressure system (e.g. Ulbrich et al., 2001) and winds associated with convection embedded in the storm field (Fink et al., 2009) contribute to the total storm intensity. Small-scale gusts often significantly exceed the mean wind speeds. They are the phenomena actually producing the damage in a windstorm field. As they are not resolved, for example, in reanalysis or forecast data, approaches for estimating their strength have been developed, basically assuming that kinetic energy from upper levels is transported downward to the surface (e.g. Brasseur, 2001; Goyette et al., 2003). The respective gust wind speeds from current regional models and parameterisations are, however, often lower than those observed, even if spatial distributions are rather well reproduced (e.g. Kunz et al., 2010).

The frequency of intense cyclones can also vary with the phase of the North Atlantic Oscillation (NAO). Donat and colleagues (2010b) demonstrated that central European windstorms are more frequent at moderately positive phases of the NAO index, which itself has an approximately normal distribution around the value of zero. Pinto and colleagues (2009) presented the influence of the NAO on the distribution of tracks of extreme cyclones, which were found to be much more organised (in a band across the North Atlantic) and extending further to the east during positive NAO phases. Still, it must be borne in mind that there are also storm occurrences during the negative NAO phase. Physically, the relationship with the NAO can be understood from the fact that a strong NAO is contributing to an enhanced westerly mean flow over the eastern Atlantic and the European continent, adding to the mostly westerly winds in the storms themselves. At the same time, the NAO is contributing to the maintenance of strong temperature gradients over the area (by cold/warm air advection from the north-west/south-west) and thus to baroclinicity. The seasonal distribution of European winter storm counts shows a maximum in January, when meridional temperature gradients are largest and suitable conditions are given to destabilise the mean zonal flow leading to baroclinic disturbances.

Recent Variability of Windstorms and Their Impacts

Changes in windstorm frequency and intensity cannot readily be estimated from the loss sums recorded by insurance and reinsurance companies. The strong increase in storm damage during recent decades may largely be attributed to inflation, increasing wealth and economic values and is further affected by changes in vulnerability. A more suitable parameter for the quantification of windstorm damage is the loss ratio, that is, the relation of total loss and total insured values. While this quantity is mostly not available for individual storm events, some accumulated numbers are regularly published. The German Insurers Association, for example, is publishing the respective numbers accumulated for the whole country every year (e.g. GDV, 2009). Several studies have been published assessing the relation of loss ratios and wind speeds based on these insurance data and on both synoptic station and reanalysis wind-speed data. One mathematical function employed assumes loss increasing with the cube of normalised wind speed in excess of the threshold of the local 98th percentile. This approach has proven to give a rather good estimation of the annual loss numbers in different regions of western-central Europe, for example in the UK (Leckebusch et al., 2007) and Germany (Donat et al., 2010a; Pearson correlation coefficient of 0.89; see Figure 11.2), in spite of shortcomings of both insurance data and meteorological data and their match (e.g. hail events included in the insurance contracts but not in the meteorological data). Clearly, neither the insurance data nor the loss numbers estimated from extreme winds over Germany show a significant trend over recent decades.

This is in line with studies by Alexandersson and colleagues (2000) and more recently by Wang and colleagues (2009) based on pressure gradients between stations over the eastern North Atlantic and north-western Europe. After homogenisation, the winter values of the extreme wind percentiles show a large decadal variability with a clear maximum in the 1990s (associated with a decade of extremely high NAO index values). Wang and colleagues (2009) demonstrated that the trends may differ for different seasons. For summer, for example, a declining trend of the percentiles is found. Based on reanalysis data for the period 1881 through 2008, Donat and colleagues (2011b) identified a significant long-term upward trend of storminess in large parts of Europe. It is related to the unprecedented high of storminess measures towards the end of the twentieth century, particularly in the North Sea and Baltic Sea regions.

11.4 Scenarios of Future Windstorms and Their Impacts

Global atmosphere-ocean models are basically able to reproduce observed storm track patterns (Ulbrich et al., 2008), cyclone distributions and numbers (e.g. Leckebusch and Ulbrich, 2004; Bengtsson et al., 2006; Leckebusch et al., 2006; Pinto et al., 2009; Ulbrich et al., 2009) and European windstorm counts and intensities (e.g. Beniston et al., 2007; Leckebusch et al., 2008a; Donat et al. 2010c). They also reproduce the observed statistical relationship between different scales. An example is the distribution of atmospheric circulation types in conjunction with European storms, though the models have a tendency of enhancing the number of zonal circulation types (Demuzere et al., 2009; Donat et al. 2010c).

Under increasing greenhouse gas concentrations in 'business-as-usual' like scenarios (IPCC SRES A1B or A2, see Nakicenovic et al., 2000), climate models produce a consistent increase of storm numbers and intensities in north-western to central Europe. This signal is, again, rather consistent between the different meteorological parameters related to storms. A significant increase of synoptic scale SLP variability in the north-east Atlantic close to western Europe was found in a multi-model ensemble of global coupled models (Ulbrich et al., 2008). A significant increase in cyclone intensity (Bengtsson et al., 2006; Della-Marta and Pinto, 2009) and the number of intense cyclones (e.g. Leckebusch et al., 2006; Pinto et al., 2007a; Leckebusch et al., 2008b) was found over north-western Europe. The statistical significance of the signal is larger when a weighting is involved which favours models with a good representation of present-day cyclone climatology (Leckebusch et al., 2008b). This signal depends, however, on the threshold of what is considered an intense cyclone. Note that the number of weaker cyclones is not significantly changing in this area, regional tendencies being dependent on the specific cyclone identification scheme used (Ulbrich and colleagues, 2013). Overall, there is a decrease in total cyclone counts in the Northern Hemisphere mid-latitudes (see Figure 11.3). Regarding extreme wind speeds, several studies analyse increased speed values over central and western Europe under future climate conditions (Pinto et al., 2007b; Gastineau and Soden, 2009; Schwierz et al., 2010; Donat et al. 2011a), in accordance with an increase in the number of storm events when using today's thresholds for the definition of those events (Donat et al., 2010c; see Figure 11.4a).

With respect to damage, winds from scenario model runs were transferred into damage by several authors. In general, an increase of damage is found for central Europe when maintaining present-day wind-loss relations and societal conditions for future climate conditions (Leckebusch et al., 2007; Pinto et al., 2007b; Schwierz et al., 2010; Donat et al., 2011a). The increases computed this way are in the order of 30 to 100 per cent for the area of Germany and the British Isles, while there is a decrease of storm risk in southern Europe (Figure 11.4b). The signals are dependent on the

(a) cyclone tracks change A1B—20C   (b) extreme cyclones change A1B—20C

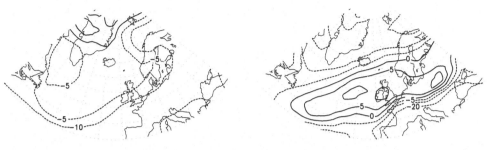

Figure 11.3 Relative change of cyclone track density in the North Atlantic/European sector in climate model simulations for the end of the twenty-first century (2071–2100) compared to recent climate (1961–2000) conditions (%). The cyclone climatologies were calculated from a multi-model ensemble of 9 GCM simulations according to the IPCC SRES A1B scenario. (a) All cyclones; (b) extreme cyclones (modified from Leckebusch et al., 2008b).

models used, and Donat and colleagues (2011a) found a significantly smaller storm damage signal from a regional model run ensemble compared to the global model ensemble investigated. It was found that even signals from different simulations with the same model can be rather large, pointing to a significant role of natural variability, so that there is still large uncertainty about the future windstorm climate in Europe, even within a specific emission scenario.

11.5 Adaptation

Buildings are subject to rules for their construction relating to wind climate. In Germany, for example, the Association of Roof Tile Producers provides rules for the construction of roofs, taking wind climate zones into account (see www.ziegeldach.de; DIN 1055–4). It can, however, be expected that there is deterioration of construction with time (e.g. a gradual loosening of anchorage of tiles), which leads to an increasing vulnerability, even though first studies of detailed damage data from insurance in Germany gave no clear evidence for such a trend. Rules for the adaptation to wind climate only apply to man-made constructions, while in nature an adaptation could be the result of a selection process. Trees not well adapted, for example, are more likely to be thrown by a storm (Bengtsson and Nilsson, 2007; Schmoeckl and Kottmeier, 2008; and references therein). Detailed studies on the adaptation strategies of natural and man-made environments to windstorm in Europe are rare (see Bolte and Degen, 2010 for a recent paper on forest adaptation).

A highly idealised approach for estimating the effects of adaptation has been used in the context of climate change effects on residential buildings, using the damage

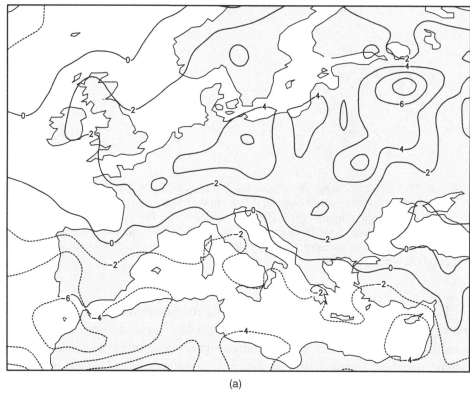

(a)

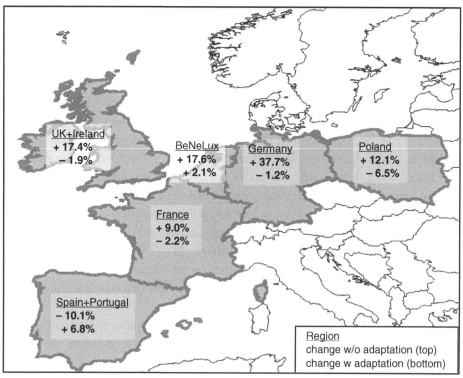

(b)

Figure 11.4 (a) Relative change of extreme wind speeds (98th percentile of daily maximum wind speeds in %) and (b) calculated losses for the end of the twenty-first century (2071–2100) compared to recent climate (1961–2000) conditions. Wind speeds and loss potentials were calculated from a multi-model ensemble of 9 GCM simulations according to the IPCC SRES A1B scenario. Loss changes are shown for each region for the cases assuming no adaptation (top row) and adaptation (bottom row) to the changing wind climate (see text for explanation). Please note that over southern Europe, adaptation to future wind conditions would mean an adaptation to weaker building structures owing to the reduced extreme wind speeds, explaining the slightly increased loss potentials in case of adaptation for the Iberian Peninsula (modified from Donat et al., 2011a).

←───

model mentioned earlier. While 'no adaptation' means calculating loss from the excess over the present day 98th percentile of wind speeds, adaptation would mean calculating loss from the future 98th percentile instead, meaning that buildings will be adapted to the changed wind climate (Leckebusch et al., 2007; Pinto et al., 2007b; Donat et al., 2011a). It turns out that this reduces the projected loss increases in central Europe drastically, even leading to lower-than-present-day loss ratios in some countries in spite of increased wind speeds. Under this assumption, changes in windstorm damage are more dependent on the occurrence of individual particularly strong events significantly exceeding the respective thresholds wind speeds.

Adaptation will also mean that the socio-economic structures have to be adapted to the occurrence of severe storm events and their potential past and future variability. Especially in the insurance and reinsurance market, a tendency can be found to base the risk management only on the most recent loss experiences. Through a temporary absence of extreme damage-related storm events, the prices in storm reinsurance decreased parallel to the surplus in reinsurance capacity after major events, for example, after the storm series in the early nineties (cf. Bresch et al., 2000). This could lead to severe consequences in the loss cover, indicating a particular risk of a too-rapid adaptation to an apparently fixed but in reality variable climate state (c.f. Changnon et al., 1997).

References

Alexandersson, H., Tuomenvirta, H., Schmith, T. and Iden, K. (2000). Trends of storms in NW Europe derived from an updated pressure data set. *Climate Research*, 14, 71–73.

Bengtsson, L. and Nilsson, C. (2007). Extreme value modelling of storm damage in Swedish forests. *Natural Hazards and Earth System Sciences*, 7, 515–521.

Bengtsson, L., Hodges, K. I. and Roeckner, E. (2006). Storm tracks and climate change. *Journal of Climate*, 19, 3518–3543.

Beniston, M., Stephenson, D. B., Christensen, O. B. et al. (2007). Future extreme events in European climate: an exploration of regional climate model projections. *Climate Change*, 81, 71–95.

Bolte, A. and Degen, B. (2010). Forest adaptation to climate change – options and limitations. *Landbauforschung Völkenrode*, 60(3), 111–118.

Brasseur, O. (2001). Development and application of a physical approach to estimating wind gusts. *Monthly Weather Review*, 129, 5–25.

Bresch, D. N., Bisping, M. and Lemcke, G. (2000). *Storm over Europe, an Underestimated Risk*. Zurich: Swiss Re Publishing. Accessed 22 October 2012 from: <http://media. swissre.com/documents/storm_over_europe_en.pdf>.

Changnon, S. A., Changnon, D., Fosse, E. R. et al. (1997). Effects of recent weather extremes on the insurance industry: major implications for the atmospheric sciences. *Bulletin of America Meteorological Society*, 78, 425–435.

Christoph, M., Ulbrich, U. and Haak, U. (1995). Faster determination of the intraseasonal variability of stormtracks using Murakami's recursive filter. *Monthly Weather Review*, 123, 578–581.

Della-Marta, P. M. and Pinto, J. G. (2009). Statistical uncertainty of changes in winter storms over the North Atlantic and Europe in an ensemble of transient climate simulations. *Geophysical Research Letters*, 36, L14703.

Della-Marta, P. M., Mathis, H., Frei, C. et al. (2009). The return period of wind storms over Europe. *International Journal of Climatology*, 29, 437–459.

Demuzere, M., Werner, M., Van Lipzig, N. P. M. and Roeckner, E. (2009). An analysis of present and future ECHAM5 pressure fields using a classification of circulation patterns. *International Journal of Climatology*, 29, 1796–1810.

Donat, M. G., Leckebusch, G. C., Wild, S. and Ulbrich, U. (2010a). Benefits and limitations of regional multi-model ensembles for storm loss estimations. *Climate Research*, 44, 211–225.

Donat, M. G., Leckebusch, G. C., Pinto, J. G. and Ulbrich, U. (2010b). Examination of wind storms over Central Europe with respect to circulation weather types and NAO phases. *International Journal of Climatology*, 30(9), 1289–1300.

Donat, M. G., Leckebusch, G. C., Pinto, J. G. and Ulbrich, U. (2010c). European storminess and associated circulation weather types: future changes deduced from a multi-model ensemble of GCM simulations. *Climate Research*, 42, 27–43.

Donat, M. G., Leckebusch, G. C., Wild, S. and Ulbrich, U. (2011a). Future changes in European winter storm losses and extreme wind speeds inferred from GCM and RCM multi-model simulations. *Natural Hazards and Earth System Sciences*, 11, 1351–1370.

Donat, M. G., Renggli, D., Wild, S. et al. (2011b). Reanalysis suggests long-term upward trends in European storminess since 1871. *Geophysical Research Letters*, 38, L14703.

Fink, A. H., Brücher, T., Ermert, V., Krüger, A. and Pinto, J. G. (2009). The European storm Kyrill in January 2007: synoptic evolution, meteorological impacts and some considerations with respect to climate change. *Natural Hazards and Earth System Sciences*, 9, 405–423.

Gastineau, G. and Soden, B. J. (2009). Model projected changes of extreme wind events in response to global warming. *Geophysical Research Letters*, 36, L10810.

GDV – Gesamtverband der Deutschen Vericherungswirtschaft (2009). *Yearbook 2009 – The German Insurance Industry*. Berlin: Gesamtverband der Deutschen Versicherungswirtschaft e.V. (German Insurance Association).

Goyette, S., Brasseur, O. and Beniston, M. (2003). Application of a new wind gust parameterisation: multiple scale studies performed with the Canadian regional climate model. *Journal of Geophysical Research*, 108, 4374.

Hofherr, T. and Kunz, M. (2010). Extreme wind climatology of winter storms in Germany. *Climate Research*, 41, 105–123.

Klawa, M. and Ulbrich, U. (2003). A model for the estimation of storm losses and the identification of severe winter storms in Germany. *Natural Hazards and Earth System Sciences*, 3, 725–732.

Kunz, M., Mohr, S., Rauthe, M., Lux, R. and Kottmeier, C. (2010). Assessment of extreme wind speeds from regional climate models – part 1: estimation of return values and their evaluation. *Natural Hazards and Earth System Sciences*, 10, 907–922.

Leckebusch, G. C., Donat, M. G., Ulbrich, U. and Pinto, J. G. (2008b). Mid-latitude cyclones and storms in an ensemble of European AOGCMs under ACC. *CLIVAR Exchanges*, 13(3), 3–5.

Leckebusch, G. C., Koffi, B., Ulbrich, U. et al. (2006). Analysis of frequency and intensity of European winter storm events from a multi-model perspective, at synoptic and regional scales. *Climate Research*, 31, 59–74.

Leckebusch, G. C., Renggli, D. and Ulbrich, U. (2008a). Development and application of an objective storm severity measure for the Northeast Atlantic region. *Meteorologische Zeitschrift*, 17(5), 575–587.

Leckebusch, G. C. and Ulbrich, U. (2004). On the relationship between cyclones and extreme windstorm events over Europe under climate change. *Global and Planetary Change*, 44(1–4), 181–193.

Leckebusch, G. C., Ulbrich, U., Fröhlich, L. and Pinto, J. G. (2007). Property loss potentials for European midlatitude storms in a changing climate. *Geophysical Research Letters*, 34, L05703.

Munich Re (1999). *Naturkatastrophen in Deutschland: Schadenerfahrungen und Schadenpotentiale (Natural Catastrophes in Germany: Damage Experiences and Damage Potentials). Order Number 2798-E-d*. Munich: Munich Re.

Munich Re (2007). *Zwischen Hoch und Tief – Wetterrisiken in Mitteleuropa, Edition Wissen (Between Highs and Lows – Weather-related Risks in Central Europe). Order Number 302–05481*. Munich: Munich Re.

Nakicenovic, N. J., Alcamo, G., Davis, B. et al. (2000). *IPCC Special Report on Emissions Scenarios*. Cambridge, UK: Cambridge University Press.

Pinto, J. G., Ulbrich, U., Leckebusch, G. C. et al. (2007a). Changes in storm track and cyclone activity in three SRES ensemble experiments with the ECHAM5/MPIOM1 GCM. *Climate Dynamics*, 29,195–210.

Pinto, J. G., Fröhlich, E. L., Leckebusch, G. C. and Ulbrich, U. (2007b). Changing European storm loss potentials under modified climate conditions according to ensemble simulations of the ECHAM5/MPI-OM1 GCM. *Natural Hazards and Earth System Sciences*, 7, 165–175.

Pinto, J. G., Zacharias, S., Fink, A. H., Leckebusch, G. C. and Ulbrich, U. (2009). Factors contributing to the development of extreme North Atlantic cyclones and their relationship with the NAO. *Climate Dynamics*, 32, 711–737.

Rockel, B. and Woth, K. (2007). Future changes in near surface wind speed extremes over Europe from an ensemble of RCM simulations. *Climate Change*, 81, 267–280.

Schmoeckl, J. and Kottmeier, C. (2008). Storm damage in the Black Forest caused by the winter storm "Lothar" – part 1: airborne damage assessment. *Natural Hazards and Earth System Sciences*, 8, 795–803.

Schwierz, C., Köllner-Heck, P., Zenklusen Mutter, E. et al. (2010). Modelling European winter wind storm losses in current and future climate. *Climate Change*, 101, 485–514.

Ulbrich, U., Fink, A., Klawa, M. and Pinto, J. G. (2001). Three extreme storms over Europe in December 1999. *Weather*, 56, 70–80.

Ulbrich, U., Leckebusch, G. C. and Pinto, J. G. (2009). Extra-tropical cyclones in the present and future climate: a review. *Theoretical and Applied Climatology*, 96(1–2), 117–131.

Ulbrich, U., Leckebusch, G. C., Grieger, J. et al. (2013). Are greenhouse gas signals of Northern Hemisphere winter extra-tropical cyclone activity dependent on the identification and tracking algorithm? *Meteorologische Zeitschrift*, 22, 61–68.

Ulbrich, U., Pinto, J. G., Kupfer, H. et al. (2008). Northern hemisphere storm tracks in an ensemble of IPCC climate change simulations. *Journal of Climate*, 21, 1669–1679.

Wang, X.-L., Zwiers, F. W., Swail, V. R. and Feng, Y. (2009). Trends and variability of storminess in the Northeast Atlantic Region, 1874–2007. *Climate Dynamics*, 33, 1179–1195.

12

The 2003 Heatwave: Impacts, Public Health Adaptation and Response in France

MATHILDE PASCAL, ALAIN LE TERTRE, AND KARINE LAAIDI

The 2003 summer in Europe was a sharp reminder that extreme temperatures remain a considerable danger for developed countries. In France, nearly 15,000 deaths were recorded during that heatwave. The most vulnerable populations were found to be elderly people; those people suffering chronic diseases, confined to bed, living alone or in social isolation; workers; and infants (Fouillet et al., 2006; Vandentorren et al., 2006; Rey et al., 2007).

Climate change will alter both the mean temperature and variance distribution. As a result, an increased frequency and duration of episodes that are today considered as extremes is expected. In the last fifty years, a significant increase in the number of warm nights has already been observed, while the increase in the occurrence of hot days is less marked (Solomon et al., 2007). The length of summer heatwaves over western Europe doubled between 1880 and 2005 (Della-Marta et al., 2007). Socio-economic deterioration, the concentration of populations in urban areas, and population ageing will lead to an increase in the number of people vulnerable to extreme temperatures. Adaptation to heat and heatwaves is therefore considered a public health priority.

Two types of adaptation to heat and heatwaves should be considered. First is a long-term task to deeply adapt society to heat through changing behaviours, improving housing and reducing the urban heat islands in cities. Second is an immediate response based on heat warning systems and heat prevention plans. Indeed, during a heatwave, most of the prevention measures are simple: To refresh, to protect oneself from sun and heat, to drink and eat regularly. However, in the absence of a pre-existing organisation, as was the case during the 2003 heatwave in France, such measures are not easily implemented, especially to assist the most vulnerable people (Vandentorren et al., 2006).

Learning lessons from the 2003 European heatwave is essential to improve future heat crisis prevention. This implies scientific studies to better characterise the environmental, social and individual risk factors necessary to orientate prevention. It also requires reflective thinking on crisis prevention and management.

12.1 Meteorological Characteristics of the 2003 Heatwave

Summer 2003 was the warmest experienced in France since the 1950s. An unusual period of hot weather started around 15 July, and a heatwave of exceptional intensity occurred between 1 and 15 August 2003, impacting the whole country. From 4 August, temperatures increased rapidly, and values above 35 °C were observed in 75 per cent of the meteorological stations and above 40 °C in 15% of the stations. Although extreme temperatures were observed in all regions, the warmest temperatures were recorded in Paris from 11 to 12 August. A later analysis of satellite data revealed contrasting day and night time heat island patterns during the heatwave, with urban heat islands of up to 8 °C greater than surrounding areas during the night (Dousset et al., 2011).

12.2 The Burden of the 2003 Heatwave

Extreme temperature can have a direct effect on the human body, leading to exhaustion, aggravation of pre-existing conditions and death. Heat-related illnesses include heat cramp, heat exhaustion and heat stroke (Batscha, 1997). However, heat strokes represent only a fraction of heat-related mortality, with heat exposure associated generally with a rapid increase in all causes of mortality. Such a pattern was observed during the 2003 heatwave, with a rapid and regular increase in rates of mortality between 4 and 13 August, followed by a progressive diminution. The daily relative risk of mortality from the heatwave increased at the beginning of August 2003, reached a peak around 12 August and then went back to usual levels around 20 August 2003 (Le Tertre et al., 2006). This escalation was observed in most cities but was especially striking in Paris (Figure 12.1). Cities located in the centre of France (Dijon, Le Mans, Lyon and especially Paris) were most impacted. At the national scale, an excess mortality of around 15,000 deaths (equivalent to a 60 per cent increase above the usual mortality levels) was observed between 1 and 20 August 2003, although the elderly suffered the greatest mortality burden of all age groups (Table 12.1).

Notably, no harvesting effect (i.e. displacement of mortality because of premature deaths during the heatwave) was observed in the following months (Le Tertre et al., 2006). This indicates that a significant gain in life expectancy would have been associated with the prevention of heat-related mortality.

Case-control studies identified major risk factors of heat-related mortality, including the lack of mobility, housing conditions, surface temperature around the house and pre-existing medical conditions (Vandentorren et al., 2006) (Table 12.2). Deprivation was also found to increase the risk, especially in the Paris area, where excess mortality

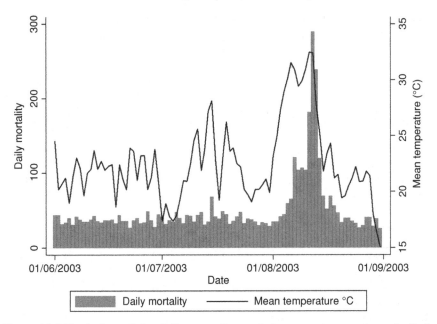

Figure 12.1 Evolution of the daily mortality and the mean temperatures in Paris during July and August 2003 (modified from Institut de Veille Sanitaire, 2003).

rates were twofold higher in the deprived areas compared to the least deprived areas (Rey et al., 2009).

12.3 The First Signals and the Initial Response

In 2003, no heat warning system existed in France, and Marseille was the only city with experience in heatwave prevention. The meteorological office Météo-France first communicated about the heatwave on 1 August 2003, pointing to the risk of drought. Occupational heat-related deaths were the first health signals recorded between 6 and 8 August. Representatives of emergency practitioners also started to report increasing difficulties, while funeral directors mentioned an unusual increase in their activities. Météo-France communicated a warning including health recommendations on

Table 12.1. *Excess mortality between 1 and 20 August 2003 (comparison with years 2000 to 2002; modified from Hémon and Jougla, 2003)*

Age (yrs)	Men	Women	Total
<44	151	−9	142
45–74	1,406	1,044	2,450
>75	3,735	8,475	1,2210
Total	**5,292**	**9,510**	**14,802**

Table 12.2. *Risk factors for death of elderly people living at home during the 2003 heatwave in France (modified from Vandentorren et al., 2006)*

	All causes	
	Odd ratio	95% CI
Occupation		
Manager	1.00	
Artisan, farmer	2.28	0.59–8.85
Intermediate occupation	1.03	0.32–3.32
Clerical and service workers	1.80	0.53–6.08
Manual workers	3.64	1.22–10.88
Social life		
Went out	1.00	
Visited cooler places	0.46	0.15–1.47
Did not leave home	2.00	0.79–5.04
Behaviour		
Dressed as usual	1.00	
Dressed lightly	0.22	0.09–0.55
Used cooling device or techniques	0.32	0.12–0.82
Dependency		
Not confined to bed, able to dress and to wash oneself	1.00	
Not confined to bed but unable to dress and to wash oneself	4.03	1.42–11.43
Confined to bed	9.59	2.89–31.79
Underlying diseases		
Cardiovascular disease	3.72	1.63–8.46
High blood pressure	1.86	0.86–4.06
Mental disorder	5.02	1.44–17.50
Neurological disease	3.52	1.04–11.98

7 August. Although a rapid investigation of the available data by the French Institute for Health Surveillance indicated that a large outbreak of mortality was occurring (Institut de Veille Sanitaire, 2003), the Ministry of Health did not order a general mobilisation in hospitals before 14 August (Lagadec, 2004).

After summer 2003, the French senate mandated a commission to review the crisis and its management (Sénat, 2004). It concluded that the risks were known but underestimated because of the unprecedented scale of the situation. The mortality response was also so rapid that it did not allow for the implementation of preventive actions.

12.4 The National Prevention Plan

Following the 2003 heatwave, the Ministry of Health initiated several actions to have a better understanding of the impacts of heat and heatwaves on mortality and morbidity in France, to organise collaboration between Météo-France and the public health authorities, to develop a prevention plan to reduce the impacts of heatwaves and to organise a real-time collection of morbidity data.

The national heatwave prevention plan involves preventive actions targeting different vulnerable populations (i.e. elderly, workers, health professionals, sportsmen). Some of the actions are enforced by law. For instance, all institutions housing elderly people or disabled people must define the organisation, the role and responsibilities of the institution during a heatwave and must give people access to a cool room at least. Note that a survey conducted in October 2003 demonstrated that 42 per cent of respondent institutions had no cooling installations available at that time (Afsset, 2004). The plan also required each city to create a database of vulnerable isolated people who should be contacted by social services during a heatwave. However, being on the list is a voluntary action, which probably excludes the most vulnerable. In addition, guidelines are available to allow different stakeholders (i.e. authorities, health professionals, non-governmental organisations [NGOs], employers) to help them establish preventive measures (Ministère de la Santé et des Sports, 2009).

During summer, brochures are disseminated through the city magazine, in the city hall, information desks, social desks, chemists and shops. They give advice on how to act during a heatwave and promote registration on the vulnerable people database. Non-governmental organisations or cities may also organise specific activities for the elderly during summer.

Heatwave warnings are issued when minimum and maximum temperatures for the next three days have a high probability of being above pre-defined thresholds (Pascal et al., 2006). During a heatwave warning, prevention messages are disseminated through media and city boards. People registered in the vulnerable registries are notified when the health warning is issued. Home visits are also organised by the city social workers, NGOs and civil servants. If necessary, hospitals and nursing home plans can be implemented to call additional staff, open more hospital beds and activate all previously planned measures.

The impact of this plan on the reduction of risks and on the excess mortality and morbidity during heatwaves is still to be determined. A key limitation to this evaluation is the lack of heatwaves since 2004. One occurred in July 2006. Minimum and maximum temperatures were slightly below those observed during the August 2003 heatwave, but July 2006 was the warmest month of July in France since 1950, and the heatwave lasted longer than that of 2003. Using a nation-wide model, researchers estimated that if the conditions had been those prevailing before 2003, 6,452 excess deaths should have been recorded during the 2006 heatwave. Yet only about 2,100 excess deaths were observed (Fouillet et al., 2008). This discrepancy may be interpreted as a decrease in the population's vulnerability to heat, which can be attributed to an increased awareness of the risk related to extreme temperatures, the implementation of preventive measures and the set-up of the warning system.

The challenge of the heat prevention plan is now to evolve in a context of increasing temperatures and individual vulnerabilities and to take into account the possible acclimatisation of the population. Temperatures have been increasing in all French departments since the 1960s, and temperatures that were previously rarely exceeded are now being recorded more and more frequently, resulting in more frequent warnings. The gradual evolution of alert thresholds cannot, however, be made unless some fundamental actions are put in place to reduce vulnerability to heat, including at times that are not alert periods, for example by working on the design of towns/cities, of housing and of transport policies (Matthies et al., 2009; O'Neill et al., 2009).

12.5 Conclusions

The 2003 heatwave was the warmest and longest ever experienced in France. Despite knowledge, mostly based on the literature, that heatwaves could be associated with severe health issues, the risk was underestimated. The population and the health professionals did not know how to protect the most vulnerable people, especially the elderly, people suffering from chronic diseases and those living alone or in social isolation. The urban population was most severely affected, the urban heat island phenomenon being found to be a major risk factor of mortality. The mortality response was so rapid that it did not allow for the implementation of preventive actions.

Since 2004, a heat prevention plan has been in place and is in continuous development. It is a good example of reactive adaptation in response to a disaster. It includes a warning system and several advices for prevention and action targeting different stakeholders. The impact of this heat prevention plan on the reduction of the risks and on the mortality and morbidity during heatwaves is still to be determined. The lower-than-expected impacts observed during the 2006 heatwaves may indicate an efficiency of the plan. Still, about 2,100 excess deaths were observed during the 2006 heatwaves, showing that heat remains a deadly risk in France.

References

Afsset – L'Agence Française de Scurité Sanitaire de L'Environnement et du Travail (2004). *Impacts Sanitaires et Energétiques des Installations de Climatisation – Établissements de Santé et Établissements Accueillant des Personnes Agées*. Maison-Alfort: L'Agence Française de Securité Sanitaire de L'Environnement et du Travail.

Batscha, C. L. (1997). Heat stroke. Keeping your clients cool in the summer. *Journal of Psychosocial Nursing and Mental Health Service*, 35(7), 12–17.

Della-Marta, M., Haylock, M., Luterbacher, J. and Wanner, H. (2007). Doubled length of western European summer heatwaves since 1880. *Journal of Geophysical Research*, 112, D15103.

Dousset, B., Gourmelon, F., Giraudeat, E. et al. (2011). *Evolution Climatique et Canicule en Milieu Urbain: Apport de la Télédétection à L'anticipation et à la Gestion de L'impact Sanitaire*. Niort: Fondation MAIF.

Fouillet, A., Rey, G., Laurent, F. et al. (2006). Excess mortality related to the August 2003 heat wave in France. *International Archives of Occupational and Environmental Health*, 80(1), 16–24.

Fouillet, A., Rey, G., Wagner, V. et al. (2008). Has the impact of heat waves on mortality changed in France since the European heat wave of summer 2003? A study of the 2006 heat wave. *International Journal of Epidemiology*, 37(2), 309–317.

Hémon, D. and Jougla, E. (2003). *Surmortalité Liée à la Canicule d'août 2003. Rapport d'étape (1/3). Estimation de la Surmortalité et Principales Caractéristiques Epidémiologiques*. Paris, France: Institut National de la Santé et de la Recherche Médicale.

Institut de Veille Sanitaire (2003). *Impact Sanitaire de la Vague de Chaleur d'août 2003 en France – Bilan et Perspectives*. Saint Maurice: Institut de Veille Sanitaire.

Lagadec, P. (2004). Understanding the French 2003 heat wave experience: beyond the heat, a multilayered challenge. *Journal of Contingencies and Crisis Management*, 12(4), 160–169.

Le Tertre, A., Lefranc, A., Eilstein, D. et al. (2006). Impact of the 2003 heatwave on all-cause mortality in 9 French cities. *Epidemiology*, 17(1), 75–79.

Matthies, F., Biokler, G., Cardenosa Marin, N. and Hales, S. (2009). *Heat-health Action Plans*. Copenhagen: World Health Organization Regional Office for Europe.

Ministère de la Santé et des Sports (2009). *Plan National Canicule – Version 2009*. Paris: Ministère de la Santé et des Sports.

O'Neill, M. S., Carter, R., Kish, J. K. et al. (2009). Preventing heat-related morbidity and mortality: new approaches in a changing climate. *Maturitas*, 64(2), 98–103.

Pascal, M., Laaidi, K., Ledrans, M. et al. (2006). France's heat health watch warning system. *International Journal of Biometeorology*, 50(3), 144–153.

Rey, G., Fouillet, A., Bessemoulin, P. et al. (2009). Heat exposure and socio-economic vulnerability as synergistic factors in heat-wave-related mortality. *European Journal of Epidemiology*, 24(9), 495–502.

Rey, G., Jougla, E., Fouillet, A. et al. (2007). The impact of major heat waves on all-cause and cause-specific mortality in France from 1971 to 2003. *International Archives of Occupational and Environmental Health*, 80(7), 615–626.

Sénat (2004). *La France et les Français Face à la Canicule: Les Leçons d'une Crise*. Paris: Sénat.

Solomon, S. D., Quin, D., Manning, M. et al. (eds.) (2007). *Climate Change 2007: The Physical Science Basis. Contribution of Working Group I to the Fourth Assessment Report of the Intergovernmental Panel on Climate Change*. Cambridge, UK: Cambridge University Press.

Vandentorren, S., Bretin, P., Zeghnoun, A. et al. (2006). August 2003 heat wave in France: risk factors for death of elderly people living at home. *European Journal of Public Health*, 16(6), 583–591.

13

Lessons from River Floods in Central Europe, 1997–2010

ZBIGNIEW W. KUNDZEWICZ

13.1 Context

In the basins of large international rivers of central Europe – the Labe/Elbe (drainage basin in Czech Republic and Germany), the Odra/Oder (drainage basin in Czech Republic, Poland and Germany) and the Vistula (most of drainage basin in Poland, with basins of tributaries located also in Slovakia, Ukraine and Belarus) – water resources are rather scarce. Even though the mean annual runoff values are low, the hydrological variability is considerable, and floods have not been uncommon. Following destructive events in the last two decades, floods have been broadly recognised as a major hazard.

13.2 Floods in Central Europe in Perspective

Two types of disastrous floods prevail in central Europe: Summer floods from mid-June to mid-August, caused by intense precipitation, and winter floods caused by snowmelt, ice advances and ice blocking. These two categories of floods have occurred many times throughout history. For example, the historical deluge, St Magdalene's flood, on 21 July 1342, caused by intense rain, devastated large areas of central Europe with thousands of fatalities, large-scale erosion (e.g. formation of deep gullies) and destruction.

The summer floods of July 1997 (Odra/Oder, Vistula and their tributaries) and August 2002 (Labe/Elbe and its tributaries) have several commonalities. First, they were generated by intensive precipitation during a longer wet spell, which covered vast areas and was caused by similar atmospheric drivers (i.e. Vb atmospheric circulation type according to the classification of von Bebber; cf. Gerstengarbe et al., 1999). Very fast, violent flash floods occurred in small and medium catchments in the mountainous tributaries of large rivers and in the upper reaches of the main rivers. Subsequently, huge masses of water travelled downstream in the main river channels, causing dike failures and inundating vast areas and large towns. The flood wave flattened as it was partially trapped in temporary storage in these areas of inundation, bringing some relief to further downstream.

Floods of 1997

The July 1997 flooding in the Odra/Oder basin caused extensive damage in three riparian countries – Czech Republic, Poland and Germany. The heavy and long-lasting rains in the beginning of July 1997 covered large areas and caused destructive flooding in the Czech Republic and Poland. From 17 to 22 July, another deluge of intensive rains was experienced, followed by a third wet spell at the end of July.

The flooding in the Upper Odra and its headwater tributaries was both violent and destructive. It destroyed the town of Kłodzko (31,000 inhabitants) located along the Nysa Kłodzka River. In Racibórz-Miedonia, the former record flooding of stage of 838 cm and discharge of 1630 m^3 s^{-1} were superseded by the much higher levels of 1,045 cm and 3,260 m^3 s^{-1}, respectively, in July 1997. Based on seven decades of records, the value of a 100-year-flood flow is estimated at 1680 m^3 s^{-1} – well below the value recorded in 1997. As the flood wave travelled downstream, alarm levels were breached continuously during the several weeks of flooding (cf. Kundzewicz et al., 1999).

The massive flood wave travelled down the Odra and inundated several river-side towns, including Racibórz (65,000 inhabitants), Opole (131,000) and Wrocław (700,000). In Opole, the water level was 173 cm higher than the absolute historic maximum (777 cm in 1997, as compared to 604 cm in 1813 and 584 cm in 1985), and the peak flow reached 3500 m^3 s^{-1}. While it was not possible to avoid inundation in these towns, the time it took for the water to travel downstream did allow some limited preparations to be made. Dike failures and massive inundations slowed the flood wave in the Odra during its downstream passage. The return period of the maximum discharge was in the order of at least several hundreds up to thousands of years in the headwaters but was far lower downstream.

The floods resulted in 114 fatalities and approximately €6.5 billion in damages. In Poland, there were 2,592 flooded towns and villages, 162,000 evacuees and around 665,000 ha of land flooded, including 450,000 ha of agricultural fields. The nation-wide economic toll in Poland for floods of summer 1997 in both the Odra and the Vistula basins was a record high.

The 1997 flood came as a surprise to the affected nations, because there had been no disastrous floods on the Odra for several decades. The 1997 deluge was the effect of exceptionally intense and persistent precipitation over a large area. This very rare hydrological event was superimposed on a complex and dynamically changing socio-economic system of the Czech Republic and Poland – countries with economies in transition – from centralised to market-based systems.

Floods of 2002

In August 2002, a Vb cyclone brought moisture from the Mediterranean area to central Europe and caused intense and long-lasting precipitation over large areas. According to Czech data, for more than a week from 6 August, every day, intense precipitation was measured in one or more Czech rain gauges. For example, extremely high precipitation (312 mm) was measured at the Cínovec station on 12 August, with high rainfall on the previous day (68 mm) and following day (26 mm). The 24-hour precipitation of 352.7 mm was recorded over a 24-hour period from 3:00 AM 12 August 2002 in Zinnwald-Georgenfeld, on the German side (Rudolf and Rapp, 2003). A new German national record for 24-hour rainfall total was set on the same day.

High precipitation caused a catastrophic flood of the river Müglitz in Germany, destroying the village Weesenstein (south of Dresden). The greatest devastation was caused by flooding on tributaries to the Elbe, such as the River Mulde hitting Grimma. The Dresden Central Station was inundated to the depth of 1.5 m by the river Weißeritz, which, during the flood, changed course back to an old riverbed. Evacuation of people and relocation of valuable mobilities was necessary, including the most valuable cultural treasures in Dresden (e.g. Zwinger Museum and Semper Opera House).

The return periods of some of the flood flows in August 2002 were on the order of several hundred years. The flood peak level in Prague exceeded all the events of the last 175 years. This is the only recorded flow rate to exceed 5,000 m^3 s^{-1}. In contrast, the flow rate of the Vltava River never reached 2,500 m^3 s^{-1} between 1941 and 2001.

The flood level of the Elbe at Dresden on 17 August 2002 was 940 cm, exceeding the former highest mark by 63 cm since the start of records in 1275. The previous highest water level (877 cm, dating back to 31 March 1845) was related to snowmelt and an ice-jam flood. Water levels in excess of 800 cm were observed in Dresden five times between 1785 and 1879 (four out of five times either in February or March) but had not been reached again since 1880. During the whole of the twentieth century, the maximum water level of the Elbe at Dresden was only 674 cm, that is, 366 cm less than the 2002 peak (Becker and Grünewald, 2003).

In the regional floods of August 2002, the damages in the basins of the rivers Labe/Elbe, Danube and their tributaries, in Czech Republic, Germany, Austria, Slovakia and Hungary totalled more than €15 billion (see Kundzewicz et al., 2005a).

Floods of 2010

Several destructive flood events occurred in central Europe in 2010 (Kundzewicz et al., 2012). In Poland, there were three major flood events – in May, June and August 2010. The 2009 to 2010 winter was snowy and cold, and there was a danger of a snowmelt *cum* ice-jam flood in the country, but fortunately the spring warming was gradual and benign, hence no flood occurred in March or April. Yet soil moisture was saturated so that intense precipitation in May (that would not normally cause a large flood) triggered massive inundations and urban flooding (e.g. in Sandomierz, a historic town on the Vistula). Although rainfall was the source of the flood waters, constrained conveyance because of improper land use (e.g. placement of the sewage treatment plant and a glass production plant) also played a role in boosting the damage in Sandomierz. Further flooding in June as the result of intense rainfall was also experienced. The death toll in Poland from these two floods was at least 25 people, with material damages estimated to exceed 10 billion PLN (more than US$3 billion). On 7 August, a flash urban flood, again caused by intense rainfall, devastated the town of Bogatynia and the surrounding areas in south-west Poland, near the borders of the Czech Republic and Germany. On 7 and 8 August, more than ten flood fatalities occurred in all three neighbouring countries. In Germany, three people died trapped in a cellar by rapidly rising flood waters. The bursting of a dam contributed to the extensive damage.

As a result of the May and June floods, Poland needed to apply for financial support from the European Union Solidarity Fund to finance the recovery and rebuild effort.

13.3 Change Detection and Attribution

Intensive and long-lasting rainfall episodes in summer, especially those induced by the Vb cyclone, as discussed previously, broke all-time records of rainfall totals and flood peak discharges and stages and ultimately led to disastrous flooding in central Europe. A climate track in floods is possible, but a broader multi-factor context must be considered.

In recent decades, an increase in the frequency and intensity of heavy precipitation events has been observed. Regional changes in timing of floods have also been observed in central Europe. There is abundant evidence of earlier occurrence of spring peak river flows and an increase in winter base flow in the basins of central Europe. Yet despite these record events in central Europe and observed changes, there is no conclusive proof as to how climate change affects flood behaviour (Kundzewicz et al., 2005b; Svensson et al., 2005).

Certainly, the occurrence of a particular flood cannot be attributed to global changes (i.e. to climate change) with a particular flood potentially being a manifest of the natural variability of the river flow process. Virtually any maximum flow rate observed recently might have been exceeded in the (possibly remote) past. However, the increased probability of intense summer precipitation and floods fits well in the general projections of the warming globe with intensified, accelerated hydrological cycles. Observed climate-related phenomena impacting on floods in central Europe include increases in precipitation intensity, increases in westerly weather patterns during winter and shrinking snow cover.

Based on climate modelling, it is projected that the expected changes in mean precipitation will differ significantly from changes in the potential occurrence of extreme rainfall events in central Europe, with the latter likely to become more intense and more frequent (Christensen and Christensen, 2003; Kundzewicz et al., 2006). To understand this difference, we consider the daily precipitation produced by the model in different intensity classes. The lower percentiles of daily precipitation over the upper part of the Odra/Oder River basin are projected to decrease slightly in the future, while the higher percentiles are likely to increase considerably. In the drainage basins of all three large rivers of central Europe (Elbe/Labe, Odra/Oder and Vistula), where dramatic floods have occurred recently, an increase in projected heavy precipitation can be noted, with potentially adverse consequences for future flood risk (Kundzewicz et al., 2006).

Hirabayashi and colleagues (2008), and Dankers and Feyen (2008) developed projections of flood hazard in Europe based on climatic and hydrological models. They project that in several major rivers in central Europe, what used to be a flood with an exceedance probability of 0.01 (i.e. 1 in 100 years or a so called 100-year flood) in the control period (1961–1990) will become considerably more frequent in further decades of the twenty-first century.

13.4 Flood Management and Flood Risk Management

Climatic impacts on water resources depend not only on changes in the characteristics of stream flows but also on such system properties as pressure or stress on the system, its management (including organisational and institutional aspects) and adaptive capacity. Climate change may challenge existing water resources management practices by contributing additional uncertainty, but in particular places, non-climatic changes may have already posed a greater impact. For example, in addition to climatic drivers, flood risk has also been increased through the transformation of flood plains adjacent to many rivers in central Europe. Large floodplain areas have been isolated from the river channels by systems of levees, and wetlands have been drained and converted to new land uses (e.g. agriculture, settlements).

The lesson from the 1997 Odra/Oder flood is that the existing flood protection system in the drainage basin, consisting of dikes, reservoirs, relief channels and a system of polders, was not designed to withstand a giant flood wave. The existing structural flood defences proved to be woefully inadequate for such a rare flood. Indeed, flood defences are designed to withstand smaller, more common floods (with return periods of decades to century) and failed when exposed to the much higher pressure of an extraordinarily high flood. Indeed, if a flood record is doubled and the flood recurrence interval gets into the range of several hundreds or thousands of years, there is no way to avoid high material losses. The event made the broader public aware of how dangerous and destructive a flood can be. It also unveiled the weaker points of the existing flood defence system and helped identify the most pressing needs for improvements.

Legislation in place in the Czech Republic and Poland was also found to be deficient. The flood occurred during a period of systemic transition from the communist regime and centrally planned economy to democracy and a market economy. The old laws were essentially abandoned and the new legal system was not yet complete and coherent, even though many new legal acts had been passed during a short time. The distribution of responsibilities was both ambiguous and conflicting. For example, as the law stood at the time of the 1997 flood in Poland, the low-level authorities were not entitled to announce either the flood alert or the alarm status. Such decisions had to come from higher authorities, being issued by the provincial anti-flood committees. As a result, these critical warnings were delayed. Local authorities typically did make common-sense decisions without waiting for instructions from above. The flow of information was also deficient, with some of the forecasts proving to be of low accuracy. The responsibilities (and cost coverage issues) were not clearly defined for the army, police and fire brigades. Polish civil defence was geared to act in case of a war rather than emergency during peace. Telecommunication support also proved vulnerable, with nearly 200,000 telecommunication links disconnected. Mobile phones provided more reliable communication, but the capacity of the system was exceeded.

Advance warning on the Odra was available for the medium and lower course after the flood occurred in headwaters in Czech Republic and Poland. As a result, the State of Brandenburg in Germany had ten days' lead time before the arrival of the flood water. However, detailed forecasts were difficult to obtain because of factors such as interruption of observations at several gauges, the flooding of the flood information office in Wrocław and so on. As a result, the need for improved forecasting and communication has been recognised. Proposed actions include building a network of weather radars; automating observations and data transmission; technical upgrading of flood warning centres, including telecommunication facilities; modernising the warning system; enhancing regional, inter-regional and international flow of flood-related information; and building more suitable forecast models.

Since the floods, considerable investments have been made. The nations and the relevant services have learned several important lessons. Indeed, when a second flood wave arrived in Poland in July 1997, the preparedness and the flood management performance were far better than those during the first crest, when the nation was largely taken by surprise. Nevertheless, overall management of the 1997 flood was rated inadequate by the Polish nation, and this assessment may have contributed to the fall of the government (see Kundzewicz et al., 1999).

During the 1997 and 2002 floods, the technical flood protection – an important element of the flood preparedness system – was found to be insufficient. Dikes have separated natural floodplains and inundation areas from the riverbed, worsening or diverting flooding. Public opinion strongly favoured the return of floodplains to rivers following the floods. Careless development of areas exposed to flooding has amplified the impact of floods. The sealing of the land surface (i.e. through development) has reduced ground storage and accelerated surface runoff. Most riparian dwellers were unaware that no technical protection can give perfect safety and that no measure prevents some inundation during extreme floods. Following the 1997, 2002 and 2010 floods, river dikes were considerably strengthened and polder areas were established.

Because of the difficulty in identifying the climate signal in the flood observation records and the large uncertainty of projections for the future (with projections being largely scenario and model dependent), no precise quantitative information can be delivered to inform flood management (Kundzewicz et al., 2010). However, in parts of Germany, flood design values have been increased by a safety margin that is based on climate change impact scenarios.

In response to the recent destructive floods in Europe and projections of a growing risk, the Floods Directive (CEC, 2007) was adopted in the European Union (EU). The directive obliges the EU Member States (therein all countries of central Europe) to undertake preliminary flood risk assessments, prepare flood hazard maps and flood risk maps (i.e. damage maps) and prepare and implement flood risk management plans. It is expected that implementation of the directive will considerably reduce the flood risk throughout central Europe.

13.5 Concluding Remarks

Having observed that flood risk and vulnerability are likely to increase in many areas of central Europe, it is important to understand the reasons for growth. These can be sought in socio-economic domain (humans encroaching into floodplain areas), terrestrial systems (land-cover changes – urbanisation, deforestation, reduction of wetlands, river regulation) and climate system. The atmospheric capacity to absorb moisture, its potential water content and thus potential for intense precipitation increases in a warmer climate.

Projections show more intense precipitation events in the warming climate for central Europe. They are of direct relevance to flood hazard in areas that have experienced recent severe floods. Although it remains debatable whether the Vb circulation track responsible for major flooding has become more frequent, it is clear that there is a growing risk of future flooding and that central Europe has considerable lessons to learn from past experiences in preparing for future events.

References

Becker, A. and Grünewald, U. (2003). Flood risk in Central Europe. *Science*, 300, 1099.
CEC – Commission of European Communities (2007). Directive 2007/60/WE of the European Parliament and of the Council of 23 October 2007 on the assessment and

management of flood risk. *Official Journal of the European Union*, 288, 27–24. Accessed 19 November 2012 from: <http://eur-lex.europa.eu/LexUriServ/LexUriServ. do?uri=OJ:L:2007:288:0027:0034:EN:PDF>.

Christensen, J. H. and Christensen, O. B. (2003). Climate modelling: severe summertime flooding in Europe. *Nature*, 421, 805–806.

Dankers, R. and Feyen, L. (2008). Climate change impact on flood hazard in Europe: an assessment based on high resolution climate simulations. *Journal of Geophysical Research*, 113, 1–17.

Gerstengarbe, F.-W., Werner P. C. and Rüge, U. (1999). *Katalog der Grosswetterlagen Europas nach P. Hess und H. Brezowski 1881–1998*, 5th edn. Potsdam/Offenbach a.M., Germany: Aufl. – Potsdam-Institut für Klimafolgenforschung/Deutscher Wetterdienst. Accessed 20 November 2012 from: <http://www.pik-potsdam.de/~uwerner/gwl/>.

Hirabayashi, Y., Kanae, S., Emori, S., Oki, T. and Kimoto, M. (2008). Global projections of changing risks of floods and droughts in a changing climate. *Hydrological Sciences Journal*, 53, 754–773.

Kundzewicz, Z. W., Dobrowolski, A., Lorenc, H. et al. (2012). Floods in Poland. In *Changes in Flood Risk in Europe*, ed. Z. W. Kundzewicz. Wallingford, Oxfordshire, UK: IAHS Press, pp. 319–334.

Kundzewicz, Z. W., Graczyk, D., Maurer, T. et al. (2005b). Trend detection in river flow series: 1. Annual maximum flow. *Hydrological Sciences Journal*, 50, 797–810.

Kundzewicz, Z. W., Lugeri, N., Dankers, R. et al. (2010). Assessing river flood risk and adaptation in Europe – review of projections for the future. *Mitigation and Adaptation Strategies for Global Change*, 15, 641–656

Kundzewicz, Z. W., Radziejewski, M. and Pińskwar, I. (2006). Precipitation extremes in the changing climate of Europe. *Climate Research*, 31, 51–58.

Kundzewicz, Z. W., Szamałek, K. and Kowalczak, P. (1999). The great flood of 1997 in Poland. *Hydrological Sciences Journal*, 44, 855–870.

Kundzewicz, Z. W., Ulbrich, U., Brücher, T. et al. (2005a). Summer floods in Central Europe – climate change track? *Natural Hazards*, 36, 165–189.

Rudolf, B. and Rapp, J. (2003). The century flood of the River Elbe in August 2002: synoptic weather development and climatological aspects. *Quarterly Report of the Operational NWP-Models of the Deutscher Wetterdienst*, 2(1), 8–23.

Svensson, C., Kundzewicz, Z. W. and Maurer, T. (2005). Trend detection in river flow series: 2. Flood and low-flow index series. *Hydrological Sciences Journal*, 50, 811–824.

14

Lessons Learned from the North Sea Flooding Disaster in the Netherlands, 1953

PIER VELLINGA AND JEROEN AERTS

14.1 What Happened?

During the night of 31 January 1953, a north-westerly windstorm hit the coastal regions around the southern parts of the North Sea in countries such as the Netherlands and the UK (see Figure 14.1). The water was pushed up against the coast by more than three metres above the normal high tide (see Figure 14.2). The conditions were exceptional, with the statistical likelihood of the water level estimated at once in 100 to 300 years. In many places, the man-made dikes along the coast were not strong enough to withstand the combination of high water and wave action. Dike failure occurred like the so-called domino effect – after the collapse of the primary coastal dike, the water entered, hit and broke through the next more inland or secondary dike. Major areas of the coastal region are at or even below sea level, so the major failure of the levee system meant rural areas, villages and towns in the area were inundated by up to three and locally more metres in a matter of hours.

The flooding disaster was unexpected by nearly all involved, and rescue efforts took several days to fully develop. Early warning systems were virtually non-existent. Many of the existing local warning mechanisms such as church bells failed largely because their use was not sufficiently ingrained in daily life and had not adequately been tested.

Housing design and construction were insufficient for large flood velocities and several metres of inundation; for example, building codes at the time did not include stability requirements under flood-water flow conditions. More than 40,000 houses were destroyed and people could not escape the water. The south-western part of the Netherlands was hit hardest with 1,835 fatalities, 100,000 evacuees, 200,000 hectares of inundated land and large economic losses of about €0.7 billion (Slager, 2010). The legal and financial responsibility of maintaining the front line of coastal defence was in the hands of the water boards directly adjacent to the sea. In practice, however, people living more inland, behind secondary dikes (former coastal dikes, now finding

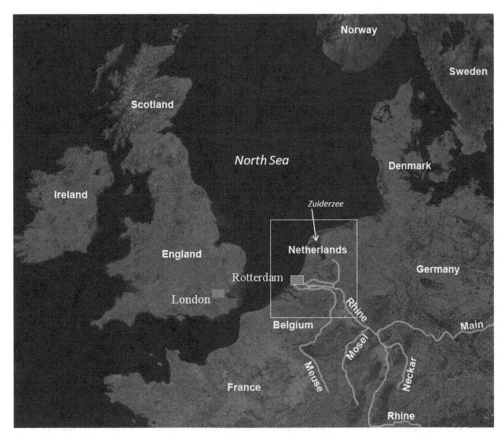

Figure 14.1 Overview of the area around the North Sea (modified from NASA Visible Earth, http://visibleearth.nasa.gov/useterms.php).

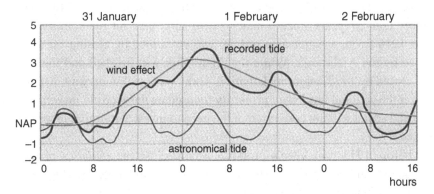

Figure 14.2 Interplay of storm surge and tidal levels above mean sea level along the North Sea coast of the Netherlands on 31 January 1953 (the Deltaworks, http://www.deltawerken.com/Deltaworks/23.html).

themselves inland because of newly reclaimed land seaward of these dikes) were equally vulnerable but politically and financially not responsible for the primary dikes. They believed that the more seaward authorities would take care of adequate protection. In practice, they were insufficiently aware of their own vulnerability.

Further casualties were avoided by the individual action of a ship's captain who moved his boat into an emerging hole in a dike that was on the verge of breaking during the height of the storm. His actions protected a much larger area of urbanised lowlands.

14.2 Really a Surprise?

In retrospect, the event should not have been a surprise. The level of maintenance in many places was below the formally adopted standards. Insufficient investment in maintenance by the Water Boards was partly a result of the damage that occurred during the Second World War and the general scarcity of funds in the years directly after the war when reconstruction of houses and infrastructure absorbed the major part of the publicly available funds. In addition, the maintenance of the primary sea defence system was in the hands of many different, relatively small water boards directly adjacent to the sea. In practice, the flooding affected a much larger inland area, not only the polders directly bordering the sea but also those lying inland.

A number of engineers of the Ministry of Public Works, Rijkswaterstaat, had evaluated the quality of the primary sea defence system in the years before the war. They concluded in 1940 that the level of safety was inadequate (Slager, 2010), but any action was superseded by the war. Directly thereafter, however, the engineer in charge of the evaluation, Mr Johan van Veen, called the attention of political leaders to the coastal vulnerability and had already developed a plan for engineering measures to reduce vulnerability (Van Veen, 1952). Before the flood, his plans were considered too ambitious and too costly to be accepted by the political system – by 1953 it had been more than 37 years since a major storm surge had hit the country. After the previous 1916 flood, a major area of land was reclaimed by closing off the estuary Zuiderzee by the Afsluitdijk. While the Afsluitdijk was finished and the Zuiderzee Estuary closed by 1932, major reclamation works were still underway. Attention and finances at the national level were primarily focused on these costly investments, with political and social will not convinced of the need to invest at a similar level elsewhere in the country.

Well before 1953, van Veen's proposal was to close the estuaries of the south-western delta (van Veen, 1952) to shorten the coastline and thereby reduce the cost of maintenance. Following the disastrous flood of 1953, van Veen's plan was largely adopted. But it was not just the losses in the south-western delta that triggered action; the narrow escape of the central part of Holland and the general feeling of vulnerability of a much larger part of the population prompted direct and far-reaching action. At the time, about four million people in the western part of the Netherlands were living at levels well below the water level experienced during the 1953 storm surge.

14.3 Vulnerability and Responsibility

The flooding took the entire population and its authorities by surprise, and they were totally unprepared for the event. Evacuation and emergency management plans were non-existent. Existing evacuation routes such as roads at a more elevated level were not effectively used. Many existing crisis measures, developed in the past, were not used because of a lack of awareness and alertness among the local authorities. At a national level, it took about three days before the full scope of the drama was clear to the political leaders and government officials.

The storm and the surge level were of such an exceptional nature that the dikes broke in more than 30 places, and secondary dikes were washed away within hours. Although in retrospect the surge water level had a statistical occurrence rate of about once every 100 to 300 years, no one had ever thought this could really happen. The focus at the time was on post-war build-up of the economy and infrastructure. The populous felt safe behind existing dikes, and there was no attention given to evacuation planning or spatial planning measures such as zoning or building codes.

Tragically, most deaths occurred more than a day after the initial flooding. When water flooded into the streets and houses, residents moved to the first floor of their houses surviving this way for a day or two, until the floors or the walls collapsed. Immediate and massive rescue by boats could have saved many people's lives. It took several days to mobilise a rescue operation because communication between those threatened people and local and national authorities had either broken down, was non-existent or was very limited.

Finally, the area covered by the local water boards responsible for maintaining the dikes in optimal condition was much smaller (only the coastal stretch) than the more inland area that was eventually flooded. In other words, there was not a good match between vulnerability and responsibility to reduce vulnerability. People in the more inland areas felt safe because they believed others (i.e. the coastal water boards) would live up to their formal mandate. In practice, the small water boards directly along the coast could not or could barely finance the works necessary to create an economically optimal safety level for the wider region.

14.4 Major Investments after the Event: The Delta Plan

After the catastrophic event of 1953, a commission was charged with researching the causes of the flood and proposing measures to prevent such disasters in the future. Johan van Veen was made the secretary of this committee. Plans were made to connect

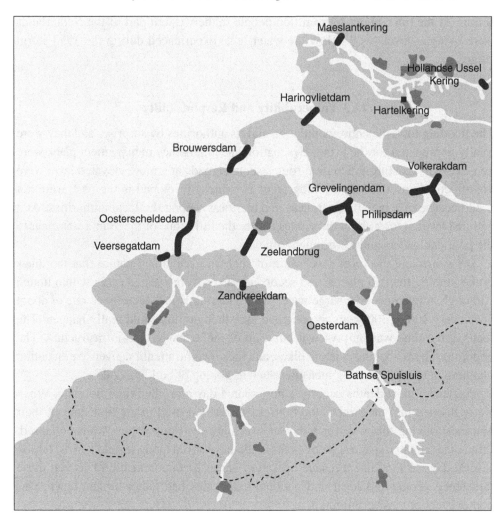

Figure 14.3 Locations of the storm surge barriers and dams of the Delta Plan that were constructed in the period 1958 and 1997 (the Deltaworks, http://www.deltawerken .com/Deltaworks/23.html).

the (semi) islands of South Holland and Zeeland with dams (the Delta Plan). The plan led to the implementation of the Deltaworks, which are a series of dams, sluices, dikes and storm surge barriers constructed between 1958 and 1997 in the south-west of the Netherlands (see Figure 14.3).

The aim of the Deltaworks was to improve flood protection by shortening the Dutch coastline, thus reducing the number of dikes that had to be raised. The length of the primary dikes in the south-west of the Netherlands was reduced with ~450 miles (~725 km). In most cases, building a barrier or a dam was much faster and cheaper than reinforcing existing dikes. Moreover, new dikes and barriers were considered safer. Some of the old dikes were founded on (geological) sand layers with potential

Table 14.1. *Costs of dams and storm surge barriers of the Deltaworks (modified from Aerts et al., 2008)*

Construction costs	Year in operation	Million Dutch Guilders	Net present value 2007 (million €)	Lifetime (years)	Design criterion Sea level rise (cm)
Storm Surge barrier Hollandse IJssel	1954	33	98		
Haringvliet barrier	1961	586	1,464	150	20
Brouwersdam	1961	141	353	100	
Hellegatsplein and Volkerak sluices	1961	191	477		
Grevelingen dam	1961	66	165		
Storm Surge barrier Oosterschelde	1986	5,500	3,850	200	20
Compartmentworks	1984	1,069	1,604		
Canal through Zuid-Beveland	1984	610	915		
Maeslant storm surge barrier	1997	990	545	100	60
Europoort barrier and Hartel barrier	1997	460	253		
Total costs of the Deltaworks		8,195	8,925		

quicksand behaviour. Such dikes could slide away if tidal channels were to migrate too close to the shore.

In the year after the flood, the Dutch government stated that plans should be made and implemented such that a major storm surge flood should never happen again. In practice, very high safety levels were adopted, based on economic reasoning. Cost-benefit analyses revealed that for the most populated areas, the dikes should be able to withstand a storm surge level with a return period of 1 in 10,000 years (van Dantzig, 1956). For less populated areas, a frequency of 1 in 4,000 was adopted. Such safety standards were covered by legislation in the National Delta Act of 1960.

The number of dams with their completion date and cost are shown in Table 14.1. It can be seen from this table that it took about fifty years to finish the works. The overall costs are estimated at about €9 billion (2007 values). The number of people protected is in the order of six million. The damage to capital goods to be avoided is more than €300 billion (Aerts and Botzen, 2011)

14.5 Evaluation of the Delta Plan

The benefits of the Delta Plan have been enormous, and not just for flood safety. Probably the major economic benefit has come from the economic development of the south-western part of the country, now protected by the dams, including industrial development, tourism development and urban development. The dams have provided access to formerly remote areas that before could only be reached via ferries or long detours. The intended benefit – no more flooding ever – has been manifested to this date, with successful protection of the area from storm surges, although events so far have not exceeded the 1953 levels.

There have been additional consequences that were not foreseen, for example in the field of ecology, water quality and sediment dynamics. A number of the originally saltwater bodies are now freshwater and, as a consequence, strongly eutrified with the green algae, making the water unfit for recreation or irrigation. Significant investment is now envisaged to turn the freshwater lakes into saltwater bodies again. Moreover, the changes in the sediment dynamics have been greater than anticipated. With every tide, sand is moving out of the estuary and not returning. Reduced tidal flows also mean the tidal flats are eroded and flattened, creating less favourable conditions for migrating birds. Again, several plans are now being developed and implemented to compensate for the losses of sand and to regenerate favourable conditions for wildlife.

Given the experience, insights and capabilities of today, it is likely that more open solutions (e.g. strengthening the existing dikes, leaving the estuaries open) would be chosen if these coastal protection plans were being made now. New techniques allow the reinforcement of old dikes, and a better understanding and appreciation of the ecological consequences of the closure of estuarine waters exists.

14.6 Climate Change and Sea Level Rise

Climate change is expected to result in higher sea levels and increases in peak level river discharge (e.g. te Linde et al., 2010). Depending on the local situation, the probability of flooding is expected to increase roughly by a factor of ten with each 50 cm rise in mean sea level (Aerts et al., 2008). However, this factor varies from place to place. For example, the largest dike ring around the cities of Rotterdam, The Hague and Amsterdam has the highest safety standard of 1 in 10,000 years. A sea level rise of 70 cm would increase the probability of flooding to approximately 1 in 1,000 years.

Climate change may also affect the frequency and pattern of windstorms, which would in turn affect storm surge levels. However, detailed model computations have shown no significant effects for the coastline of the Netherlands (Sterl et al., 2008; 2009). The maximum (storm) wind force is expected to increase, but simultaneously the (storm) wind directions are expected to change such that the piling up of water effect would be less along the coast of the Netherlands.

In 2008, a special commission developed the second Delta Plan, addressing long-term spatial planning and water management measures for the Netherlands in the face of climate change (Deltacommissie, 2008). This committee concluded that the spatial contours of the Netherlands could be maintained even under conditions of 2 to 3 m of

Table 14.2. *Cost of maintaining flood protection standards under conditions of climate change*

	Scenarios				
Year	2040	2100	2100	2100	Far future
Sea level rise (cm)	24 cm	60 cm	85 cm	150 cm	500 cm
Max discharge river Rhine (m^3/s)	16,700	18,000	18,000	18,000	18,000
Max discharge river Meuse (m^3/s)	4,200	4,600	4,600	4,600	4,600
River works	Costs (billion €)				
River widening Rhine	2.7	5.5	5.5	5.5	5.5
River widening Meuse	1.3	4.2	4.2	4.2	4.2
Dike reinforcement	0.2	1.8	2.6	6.1	36
Coast	Costs (billion €)				
Beach nourishment Holland	1.9	6.4	9.1	16.0	25
Beach nourishment Waddensea	1.1	3.8	5.4	9.6	Unknown
Beach nourishment Westerschelde	0.1	0.4	0.6	1.1	Unknown
Coastal dike reinforcement	1.9	2.3	2.6	3.4	8
Total	9	24	30	46	80

Source: modified from Aerts et al. (2008).

sea level rise. The committee advised to keep the estuary open such that river water could continue to flow freely into the sea. The overall cost of maintaining an adequate level of flood protection would be in the order of 0.2 to 0.4 per cent of the gross domestic product (GDP). Table 14.2 specifies the cost of additional flood protection measures under different climate change scenarios. The total costs include the flood protection works for 3,500 km of river, estuary and sea dikes, the nourishment of 450 km of beaches and broadening of the main rivers (Table 14.2).

The new Delta Committee was particularly interested in high-end scenarios for climate change, its objective being that even with the most extreme sea level rise, spatial investments in urban development and infrastructure should be safe. The high-end scenario for sea level rise (Katsman et al., 2011) is shown in Figure 14.4, together with the more short-term scenarios developed by the Netherlands Meteorological Office (KNMI, van den Hurk et al., 2007; see also the discussion by Nicholls et al., 2011). The high-end scenario serves as an ultimate 'what-if' check for any plan to be developed. For short-term operational planning purposes, a range of scenarios is considered.

14.7 Flooding Disasters: A Historic Necessity for Improvements in Flood Protection?

The history of the Netherlands illustrates that a major flooding disaster is often the necessary catalyst for taking economically cost-effective measures to improve the

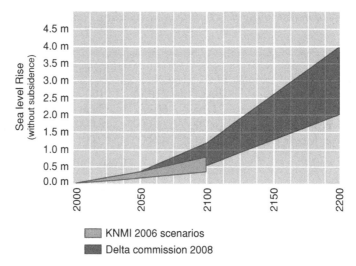

Figure 14.4 Sea level rise scenarios, including upper-end scenarios developed for the New Delta Committee (Deltacommissie, 2008).

protections of low-lying areas against flooding. This observation is supported by events in New Orleans (hurricane Katrina and follow-up activities) and Japan (the 2011 tsunami and its follow-up activities). But can this sequence of tragedy and remedy be broken? The second Delta Plan initiated in the Netherlands in 2008 is an attempt to do so and has had two important drivers. One is the events surrounding hurricane Katrina in New Orleans. The flooding of New Orleans and the loss of life and property triggered a national (political) debate on safety against flooding in the Netherlands. The second driver is climate change and the acceleration of sea level rise and the parallel notion that the historic statistics of storm surges may no longer be a reliable basis for setting safety standards.

One of the more important elements of the second Delta Plan is the proposed increase of the safety standard. The new standard now under consideration includes both the frequency of specific storm surge levels that dikes should be able to withstand and the measures to be taken to reduce damage in the case of water breaching coastal protection works. The concept of 'unbreachable' dikes is being explored. During an extreme storm surge, the seawater may wash over the dike, but the dike would be so large and/or solid that it would not break (Vellinga et al., 2009).

Low-lying regions worldwide are particularly vulnerable to floods. This is generally understood by experts but often not appreciated by the people that live in these areas, with major flooding continuing to take residents by surprise. Undoubtedly this is a result of the relatively low frequency of events. In practise, however, it is relatively easy to determine an economically efficient safety level. Many low-lying urban centres seem to be 'under-protected' and lack the investments in flood management measures to protect people and assets. Scientific analyses can help to create awareness about the actual flood safety conditions and what can be done to improve the level of safety if considered economic and necessary.

References

Aerts, J. C. J. H., Bannink, B. and Sprong, T. A. (eds.) (2008). *Aandacht voor Veiligheid*. Amsterdam: VU University Press.

Aerts, J. C. J. H. and Botzen, W. J. (2011). Climate change impacts on long-term flood risk and insurance: a comprehensive study for the Netherlands. *Global Environmental Change*, 21, 1045–1060.

Deltacommissie (2008). *Working Together with Water*. The Hague: Deltacommissie. Accessed 24 October 2012 from: <http://www.deltacommissie.com/advies>.

Katsman, C. A., Sterl, A., Beersma, J. J. et al. (2011). Exploring high-end scenarios for local sea level rise to develop flood protection strategies for a low-lying delta – the Netherlands as an example. *Climatic Change*, 109, 617–645.

Nicholls, R. J., Marinova, N., Lowe, J. A. et al. (2011). Sea-level rise and its possible impacts given a 'beyond 4 degree world' in the 21st century. *Philosophical Transactions of the Royal Society*, 369, 161–181.

Slager, K. (2010). *Watersnood*. Gouda, Netherlands: De Buitenspelers Publishers in cooperation with the Watersnood Museum.

Sterl, A., van den Brink, H. W., de Vries, H., Haarsma, R. and van Meijgaard, E. (2009). An ensemble study of extreme North Sea storm surges in a changing climate. *Ocean Science*, 5, 369–378.

Sterl, A., Weisse, R., Lowe, J. and von Storch, H. (2008). Winds and storm surges along the Dutch coast. In *Exploring High-end Climate Change Scenarios for Flood Protection of the Netherlands*, ed. P. Vellinga. De Bilt: KNMI/Alterra, pp. 82–98. Accessed 24 October 2012 from: <http://www.knmi.nl/bibliotheek/knmipubWR/WR2009-05.pdf>.

te Linde, A. H., Aerts, J. C. J. H., Bakker, A. M. R. and Kwadijk, J. C. J. (2010). Simulating low-probability peak discharges for the Rhine basin using resampled climate modeling data. *Water Resources Research*, 46, 1–19.

van Dantzig, D. (1956). Economic decision problems for flood prevention. *Econometrica*, 24, 276–287.

van den Hurk, B., Klein Tank, A., Lenderink, G. et al. (2007). New climate change scenarios for the Netherlands. *Water Science and Technology*, 56, 27–33.

van Veen, J. (1952). *Dredge Drain Reclaim: The Art of a Nation*, 3rd edn. The Hague, the Netherlands: Martinus Nijhoff Publishers.

Vellinga, P., Marinova, N. and Loon-Steensma, J. M. V. (2009). Adaptation to climate change: a framework for analysis with examples from the Netherlands. *Built Environment*, 35, 452–470.

Part IV

Case Studies from the Developing World

15

Adapting to Drought in the West African Sahel

SIMON P. J. BATTERBURY AND MICHAEL J. MORTIMORE

Local responses to drought in the West African Sahel tell a remarkable story, in which local adaptive capacity has proven equal to the challenges of living in a harsh and drought-prone environment. This chapter outlines the extent of this capacity and argues that it should be facilitated and celebrated.

The Western Sahel is a vast and populous dryland region running east-west between the Sahara Desert to the north and the West African forest zone to the south. A 'band' of broadly similar climatic conditions runs from Senegal to the shores of Lake Chad. Travelling north from the West African coast, rainfall diminishes and average biomass decreases. Despite its remoteness, population density is moderate and high around urban centres. The region has been traversed by armies, traders and new settlers for thousands of years. There is a diversity of cultures, languages, political entities and production systems. Rural livelihood systems are based on rainfed agriculture and pastoralism, always influenced by climatic hazards: Drought and temperature extremes and occasional flooding and high winds. The rural heartland also produces food for markets and growing urban areas, the largest in the drylands being the cities of Dakar, Bamako, Ouagadougou, Niamey and Kano.

Dryland adaptation is deep seated among Sahelian people. Food shortages, occasioned by insufficient precipitation to nourish crops, sometimes extend into famine after extremely dry years, and these are common over the long-term record of human occupation. It is important to understand the concrete responses that individuals and households make when faced with seasonal and longer-term climatic risks (Mortimore, 1989). Negotiating the drought has been pursued outside the public gaze – often unceremoniously – through what anthropologist Paul Richards calls 'unsupervised learning' at the local scale (Richards, 2010: 18) and through the transmission of learned practices to others. Local people have found ways to minimise the risk of lost food production and have initiated a gradual and sometimes vibrant expansion of commercial activity alongside increasing reliance on remittances from migration. Their strategies also involve circumventing the disruptive effects of the expropriation

149

of productive assets by governments and some private actors. On occasion, outside knowledge has blended well with local expertise to enhance agricultural resilience. For example, animal health protection and provision of water points are present in a few drier pastoral areas and have proven valuable unless over-exploited.

15.1 The Sahel – Physical Characteristics and Drought

Over millennia, the Sahel has seen several phases of drying and wetting, now thought to be linked to the strength of the Atlantic Meridional Overturning Circulation. These phases have in turn influenced human settlement and migration, with a population exodus during drier periods (Mulitza et al. 2008; Castañeda et al., 2009). Aridity and rainfall variability increase with latitude. In the absence of irrigation, crop production is limited to where average annual rainfall is more than 250 to 300 mm (Gubbels, 2011). North of this is the pastoral zone, with seasonal grazing possible in drier conditions.

The rainfall trend over the past century has exhibited relatively abrupt changes in direction, with a multi-decadal wet period (1951–1968) followed by a severe drying period from the late 1960s to the 1990s. There has since been a rainfall recovery but with further dry years (Figure 12.1). The decreasing rainfall and intensity of the droughts in the Sahel from the 1970s to the 1990s and the extent of the downward trend in rainfall are the most significant recent climate changes recognised by climatologists (Dai et al., 2004).

In particular, unreliability has characterised the last forty years, with a punishing drought from 1972 to 1974. There was a further, deeper dry period in the mid-1980s for which farmers and governments were better prepared. From the mid-1960s to the mid-1990s, annual rainfall declined by 25 to 30 per cent depending on location (Figure 15.1). The east-west isohyets of mean annual rainfall migrated southwards by up to 100 km (Mortimore, 2010). From 1968, growing seasons were significantly shortened by late starts and early finishes.

Future climatic change in the Sahel is uncertain, with some disagreement among climate models. Models do not agree on the sign (direction) of change in future – while some predict a persistently dry regime, others predict an improvement (Giannini et al., 2008).

15.2 Responses to Drought in Northern Nigeria in the 1970s

Northern Nigeria provides the evidence that Sahelian societies adapt to drought hazard over time, and it is a region in which little national or international aid has assisted this process. The drought-prone region now has 30 to 40 million people (Mortimore, 2010). Farming and herding include significant commercial production. In the 1960s when rains were good, the region saw booms in commercial groundnut and cotton, and this diverted labour and land from food crops, as it had in 1914 and 1927 (Watts,

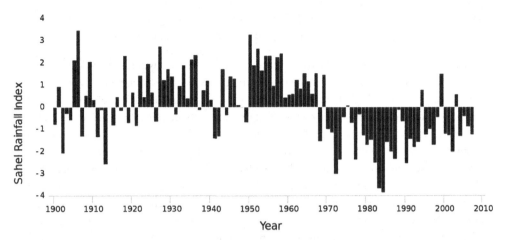

Figure 15.1 Long-term Sahel rainfall trends (Wikimedia Commons, Accessed 11 June 2013 from <http://commons.wikimedia.org/wiki/File:Sahel_rainfall_timeseries_en. svg>).

1983). The 1972 to 1974 drought was punishing, and the plight of rural households was real. Granaries were emptied, sometimes for years. People ate the leaves of trees and herbs or sold livestock and other valued possessions to buy cassava flour at markets as a substitute for grain (Mortimore, 1989; Mortimore and Adams, 2001). Some families went to the towns to beg, especially livestock herders. There was widespread undernutrition and very high livestock mortality. The government seemed powerless or unwilling to help. In 1974, the Nigerian states distributed grain, enough to offer a few meals for the luckier families; stories of corruption and misallocation abounded. Ironically, food aid was glimpsed passing by on trucks in transit to Niger to the north, where the international media had focused world attention.

As the severity of the food shortage deepened, individuals and households adjusted on a trajectory of diminishing reversibility. Offering labour to others for food or cash, gathering famine foods or getting help from kin gradually gave way in time to selling livestock, pledging or selling land and, eventually, begging and migrating as a family to more promising regions. The routine use of 'famine foods' was extended dramatically (Mortimore, 1989: 67–74).

Given the situation that prevailed, is it correct to paint the drought cycle of 1972 to 1974 as evidence of the failure of adaptive strategies? Or were these strategies temporarily overwhelmed by the severity of the drought, yet they remained resilient? Did people learn from their severe hardships?

15.3 From Coping to Adaptation Strategies

The evidence is that short-term coping feeds into enduring adaptive strategies for managing variable rainfall and bio-productivity. Central to this is reduced reliance on rainfall (Batterbury and Forsyth, 1999). This was particularly important when very severe Sahel drought returned from 1982 to 1984 (Figure 15.1). A basic model suggests a sequence of stepwise diversification strategies is followed:

1. Farmers *negotiate* rainfall each year. Pearl millet is the most drought resistant of all the crops grown in the Sahel. Farmers manage their own gene pool by selecting and storing the best seed (Busso et al., 2000). Diversity in landraces is a recognised defence against disease, pests and climatic hazards. Cropping patterns and labour allocation are finely judged according to rainfall, particularly for weeding. In some wetter areas, productivity rises with population density (rather than the reverse) such that families now practice sustainable intensification on tiny plots (at less than 0.5 ha per capita) (Adams and Mortimore, 1997). What matters most – more than rainfall – is a farmer's ability to mobilise enough labour to meet peak requirements and to access enough organic material to sustain soil biological health (Mortimore, 2006).

2. For farmers and herders alike, *livestock* are a repository for savings, a reserve for contingencies, a self-reproducing asset, and a source of income and energy. Livestock herding is less sensitive than crop production to climatic changes (Mertz et al., 2010). Animals also support intensification on the farm (through the cycling of nutrients through crop residues and manure, although human labour input is high when animals are stall fed). Animals are bred for their drought and disease tolerance and grazing performance under variable conditions (Kratli, 2008). Over the last thirty years, there has been a shift to small ruminants (sheep and goats) across the Sahel, as they are less costly, hardier, easier to feed and reproduce more quickly than cattle (de Leeuw et al., 1995).

3. *Business.* To minimise risk from crop failure and to tackle poverty, alternative livelihoods and sources of income are necessary and desirable. Almost all individuals in four Nigerian villages surveyed in the mid-1990s had income from trading, making articles for sale or providing services (Mortimore and Adams, 2001). Incomes may be small, but they are important in the pre-harvest season when granaries are depleted. Women are increasingly turning to small-scale business, including food sales and small-scale trading, as a study of Zarma households in Niger showed (Batterbury, 2001). Onions, tomatoes and mangoes are increasingly grown for sale to urban centres, which continue to grow in size (Sahel Working Group, 2007; Gubbels, 20011).

4. Greater degrees of *livelihood diversification* include non-local activities. They are associated with high levels of mobility among rural communities, towns and cities and commercial farms. As part of diversifying away from subsistence activities, human mobility has increased with the high rates of urbanisation and better transport (modern vehicles and more surfaced roads). This does not indicate the failure of rural livelihoods but rather their success; drought requires a diversified portfolio of activities (Figure 15.2). The exception is 'distress migration' in the worst years of food insecurity in the 1970s, when thousands of single men and some families moved south away from the drought in search of work.

15.4 Productive Bricolage

Stepwise diversification of livelihood activities (Figure 15.2) occurs as drought conditions worsen. Nonetheless, in less difficult years, it is actually easier for households to

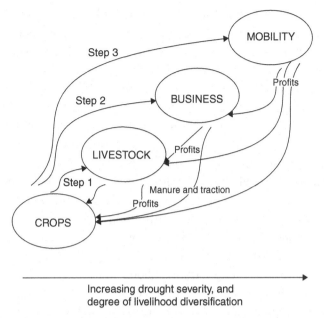

Figure 15.2 Typical stepwise diversification of livelihood activities in a rainfed agricultural community in the Sahel. Frequently, multiple activities are practiced in a household, and assets are pooled. Step 1 – combine crops and livestock, Step 2 – add local household business activity, Step 3 – some household members migrate to seek work and remittances (Adams and Mortimore, 1997).

pursue multiple strategies at the same time, and this process becomes less linear and more complex. Tailoring adaptive drought management to circumstances therefore involves *productive bricolage*, the controlled mixing of income sources and subsistence activities (Batterbury, 2001). To operate these strategies requires astute management of scarce capital and an institutional space that permits negotiation amongst household members. This assures scarce labour demands are met and opportunities are exploited.

New technology and infrastructure can assist *bricolage*. Across the Sahel, mobile phones – not present in the 1970s and 1980s – facilitate communication between household members, particularly over price information for sales and purchases, and labour management. Seasonal climate forecasting is gaining in accuracy but is still not available to the majority of Sahelian farmers and herders (Tacko Kandji et al., 2006).

At the same time, political and economic forces confound *productive bricolage*. Conflict in Côte d'Ivoire, for example, has curtailed a favoured migration route for young Burkinabè men. In the 1990s, their remittances balanced the Burkina Faso economy, being 6 per cent of GDP. The response to blocked labour migration routes is usually swift. Because of the conflict, travel increased to Libya's oil towns. The European Union tightened its borders to deny travel further northward than Morocco, increasing movement to Libya (Bredeloup and Pliez, 2011). Political uncertainty in 2011 has ceased this livelihood opportunity but local gold mines in the Sahel attract migrants. Similarly, oil-rich central Africa currently hosts growing numbers of temporary Sahelian workers in Gabon and neighbouring countries.

There are also unwelcome losses of good agricultural land to agri-food development for external markets on the few large irrigation schemes in the Sahel. Urban and foreign buyers are currently securing large holdings in more productive areas as investments, although these are not as extensive as in eastern and southern Africa and some production is destined to supply local urban areas (Hilhorst et al., 2011).

It seems certain that Sahelian households will need more opportunities for diversification in future, as even the coastal and north African migration destinations have become more constrained. This is particularly true for populations in Burkina Faso, Niger and Mali that have historically relied on long-term migration to the labour markets of the francophone West African coast. Such alternatives need to be understood as components of integrated, long-term livelihood strategies.

Returning to the questions posed at the end of section 15.3, the dire assessment of Sahelian futures that followed the 1970s famines was in fact an under-estimation of 'adaptive success' across the Sahel. In fact, few mortalities could be attributed directly to the Sahel drought, and there were fewer in the 1980s. Since that time, *bricolage* activities have expanded.

15.5 Climate and Development Aid

Mortimore (2010) refers to northern Nigeria, where development aid has been limited and drought adaptation has occurred largely without outside support, when he says: 'Some 25 years after the Sahel drought seemed to question the very survival of human communities, the farming systems persisted, village populations were stable or increasing, farming and livestock production were integrating and intensifying, and livelihoods were diversifying. Until 2008, visible indicators of monetisation and investment were increasing' (Mortimore, 2010: 136).

In contrast, the francophone Sahel has received millions of dollars in aid to increase its resilience to drought since the mid-1970s. Some of this has gone to poor rural communities, ostensibly to reduce vulnerability and alleviate poverty. However, one assessment placed this as only US$45 per person per year by the mid-2000s (Sahel Working Group, 2007: 28). Well-targeted development aid can improve resilience to drought and should not be dismissed as an inefficient type of transfer payment (Mortimore, 2010). The restoration of thousands of hectares of remote village farmland in Burkina Faso through soil and water conservation techniques, built by farmers but with some outside logistical and technical support from Oxfam and German and Dutch aid, was positive (Batterbury, 1998). Burkinabè farmers in the drylands of that country have often exploited such project aid to gain additional status and benefits in ways unintended by donors (e.g. to attract a gift of a school or a borehole to a village that was performing well with an existing soil restoration donor), but nonetheless their resilience has been enhanced (Batterbury, 1998).

Paul Richards has argued, following Emile Durkheim's work, that 'much of what farmers and farm labourers do (as opposed to what science or regulatory authority tells them to do) is based on practical contingencies... or it relates to unavoidable social responsibilities to dependants or a wider community group' (Richards, 2010: 10).

If we make these practical contingencies our starting point in understanding adaptation to climate change rather than dismissing them as incidental or inefficient, then we throw up an enduring challenge to climate and development policy. We need to recognise structure and agency – by showing sensitivity to the internal adaptive resources of farming and pastoral households as well as recognising the constraints they face and understanding their autonomous capabilities, as well as the external forces bearing upon them.

Meanwhile, the international aid sector could place much greater attention on the constraints that Sahelian peoples cannot themselves address under climate change. These include international migration restrictions and threats to productive activity like cheap food imports to urban centres, unjust land alienation and weak human and animal health services.

15.6 Conclusion

In conclusion, we have argued that the severity of droughts affecting the Sahel has led to practical adaptive strategies emerging, as well as long-term adjustments of rural livelihood activities to rainfall deficiencies and uncertainties. Forty years of drought adaptation in the Sahel have been overlooked in contemporary debates over 'adaptation to climate change' that focus most attentively on large-scale, public-sector adaptation planning and unjustly polarise regional responses into 'mitigation' or 'adaptation'. Adaptation is not just an issue for the public sector to tackle and for planners to organise on behalf of local people. It still includes direct responses to climate hazards by poorer members of rural society, with an attendant growth of local knowledge about these hazards. The seriousness of African drought and famine is often noted in the climate change literature, but long-term adaptive strategising and learning from hardship have too often been misinterpreted as mere 'coping strategies'.

Sahelian adaptation is much more than this. It has resulted from a combination of skill and technique, both learned and invented, and has been combined with deliberate efforts by households and communities to reduce their vulnerability. There is strong evidence that the response to Sahelian drought over the last forty years, while constrained, is now adequate to the challenge of future climate uncertainty. Adaptation should be facilitated rather than driven from the outside.

Acknowledgements

Our findings come from residence and field research from 1963 to 1986 in northern Nigeria (MM), subsequent research projects funded by the ESRC, DfID, World Bank and SSRC (MM and SB) including three years' fieldwork in Burkina Faso and Niger (SB). We acknowledge the comments of two referees.

References

Adams, W. M. and Mortimore, M. J. (1997). Agricultural intensification and flexibility in the Nigerian Sahel. *The Geographical Journal*, 163(2), 150–160.

Batterbury, S. P. J. (1998). Local environmental management, land degradation and the 'Gestion des Terroirs' approach in West Africa; policies and pitfalls. *Journal of International Development*, 10, 871–898.

Batterbury, S. P. J. (2001). Landscapes of diversity: a local political ecology of livelihood diversification in south-western Niger. *Ecumene*, 8, 437–464.

Batterbury, S. P. J. and Forsyth, T. J. (1999). Fighting back: human adaptations in marginal environments. *Environment*, 41(6), 6–11.

Bredeloup, S. and Pliez, O. (2011). *The Libyan Migration Corridor*. San Domenico di Fiesole, Italy: Robert Schuman Centre for Advanced Studies, European University Institute. Accessed 24 October 2012 from: <http://www.eui.eu/Projects/ TransatlanticProject/Documents/CaseStudies/EU-USImmigrationSystems- Security-CS.pdf>.

Busso C. S., Devos, K. M., Ross, G. et al. (2000). Genetic diversity within and among landraces of pearl millet (*Pennisetum glaucum*) under farmer management in West Africa. *Genetic Resources and Crop Evolution*, 47, 561–568.

Castañeda, I. S., Mulitza, S., Schefuss, E. et al. (2009). Wet phases in the Sahara/Sahel region and human migration patterns in North Africa. *Proceedings of the National Academy of Sciences of the United States of America*, 106, 20159–20163.

Dai, A., Lamb, P. J., Trenberth, K. E. et al. (2004). The Sahel drought is real. *International Journal of Climatology*, 24, 1323–1331.

de Leeuw, P. N., Reynolds, L. and Rey, B. (1995). Nutrient transfers in West African agricultural systems. In *Livestock and Sustainable Nutrient Cycling in Mixed Farming Systems of Sub-Saharan Africa. Volume II: Technical Papers. Proceedings of an International Conference held in Addis Ababa, Ethiopia, 22–26 November 1993*, eds. J. M. Powell, S. Fernandez-Rivera, T. O. Williams and C. Renard. Addis Ababa: International Livestock Centre for Africa, pp. 371–392.

Giannini, A., Biasutti, M. and Verstraete, M. M. (2008). A climate model-based review of drought in the Sahel: desertification, the re-greening and climate change. *Global and Planetary Change*, 64, 119–128.

Gubbels, P. (2011). *Escaping the Hunger Cycle: Pathways to Resilience in the Sahel*. Washington, DC: Groundswell International. Accessed 29 October 2012 from: <http:// www.e-alliance.ch/fileadmin/user_upload/docs/Publications/Food/2012/Escaping_the_ Hunger_Cycle_English.pdf>.

Hilhorst, T., Nelen, J. and Traoré, N. (2011). Agrarian change under the radar screen: rising farmland acquisitions by domestic investors. Paper presented at *International Conference on Global Land Grabbing*, Institute of Development Studies, Sussex, UK, 6–8 April.

Kratli, S. (2008). *Time to Outbreed Animal Science? A Cattle Breeding System Exploiting Structural Unpredictability. The WoDaaBe Herders in Niger. STEPS Working Paper No. 7*. Brighton, UK: Institute of Development Studies.

Mertz, O., Mbow, C., Østergaard Nielsen, J. et al. (2010). Climate factors play a limited role for past adaptation strategies in West Africa. *Ecology and Society*, 15, 25.

Mortimore, M. J. (1989). *Adapting to Drought: Farmers, Famines and Desertification in West Africa*. Cambridge, UK: Cambridge University Press.

Mortimore, M. J. (2006). Managing soil fertility on small family farms in African Drylands. In *Biological Approaches to Sustainable Soil Systems*, eds. N. Uphoff, A. S. Ball, E. Fernandes et al. New York: Taylor and Francis, pp. 373–390.

Mortimore, M. J. (2010). Adapting to drought in the Sahel: lessons for climate change. *Wiley Interdisciplinary Reviews: Climate Change*, 1, 134–143.

Mortimore, M. J. and Adams, W. M. (2001). Farmer adaptation, change and 'crisis' in the Sahel. *Global Environmental Change*, 11, 49–57.

Mulitza, S., Prange, M., Stuut, J.-B. et al. (2008). Sahel megadroughts triggered by glacial slowdowns of Atlantic meridional overturning. *Paleoceanography*, 23, 1–11.

Richards, P. (2010). *A Green Revolution from Below? Science and Technology for Global Food Security and Poverty Alleviation*. Wageningen, Netherlands: Wageningen University. Accessed 24 October 2012 from: <http://edepot.wur.nl/165231>.

Sahel Working Group (2007). *Beyond Any Drought. Root causes of chronic vulnerability in the Sahel*. Accessed 11 June 2013 from: <http://reliefweb.int/sites/reliefweb.int/files/resources/E3B50A8AB7395B02C12573A2004A8428-swg_june2007.pdf >

Tacko Kandji, S., Verchot, L. and Mackensen, J. (2006). *Climate Change and Variability in the Sahel Region: Impacts and Adaptation Strategies in the Agricultural Sector*. Nairobi, Kenya: World Agroforestry Centre and United Nations Environment Programme.

Watts, M. J. (1983). *Silent Violence: Food, Famine and the Peasantry in Northern Nigeria*. Berkeley: University of California Press.

16

The 2004 Indian Ocean Tsunami: Sri Lankan Experience

SAM S. L. HETTIARACHCHI AND W. PRIYAN S. DIAS

16.1 The Event

The Indian Ocean tsunami of 26 December 2004 affected many countries around the rim of the Indian Ocean, particularly Indonesia, and to a lesser but considerable extent Sri Lanka. Although this paper focuses on the Sri Lankan experience, the underlying or emergent principles have wide applicability. The tsunami arose from a massive submarine earthquake 400 km west of northern Sumatra at around 6:30 AM Sri Lanka time. The earthquake measured 9.3 on the Richter Scale and the fault length exceeded 1,000 km. The tsunami reached Sri Lanka's east coast around 8:30 AM and its west coast around 9:30 AM. More than two-thirds of Sri Lanka's 1,400 km coastline was affected.

The previous tsunami to have affected Sri Lanka occurred on 27 August 1883, arising from an eruption of the volcanic island of Krakatoa. On that occasion, unusually high water levels were observed in several areas along the eastern and southern coastline, though with hardly any damage. Given that more than a century had elapsed since this previous event, itself relatively small, public awareness regarding tsunamis was virtually non-existent. If at all, people in Sri Lanka associated tsunamis with the Pacific rather than the Indian Ocean. The majority would not have realised that the event was a tsunami even when it was unfolding. There was no infrastructure for warning in real time, although two to three hours elapsed before the waves struck Sri Lanka's shores after the tsunamigenic earthquake, with an hour in between the arrival times on the east and west coasts.

A particular aspect of Sri Lanka's vulnerability is that economic development, population density and transport infrastructure are concentrated along the coast. As a result, around 40,000 people died and around 120,000 houses were destroyed or damaged by the tsunami. The direct economic loss relating to assets and infrastructure was close to US$1 billion. The asset loss was largely in the housing sector and estimated at US$306 to 341 million (ADB, 2005). This can be attributed to the

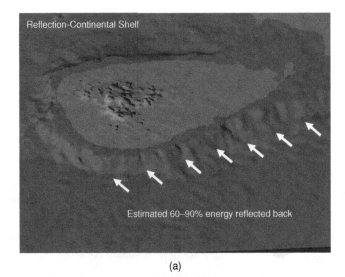

Figure 16.1 (a) Reflection and (b) transformation from the continental shelf (Hettiarachchi and Samarawickrama, 2005).

location of residential buildings close to the coastal transport routes. For instance, 54 per cent of the damaged housing units were located within 100 m of the shoreline on the densely populated south-western coast (DCS, 2007).

Being a small island state, Sri Lanka's perimeter-to-area ratio is relatively high, and most of its coastal perimeter was affected by direct, refracted, diffracted and reflected tsunami waves. The island has a very narrow continental shelf with a sudden drop in level from around 150 to 200 m to 3,000 m. A portion of the incoming wave energy may have been reflected from this continental shelf (Figure 16.1a). The wave energy

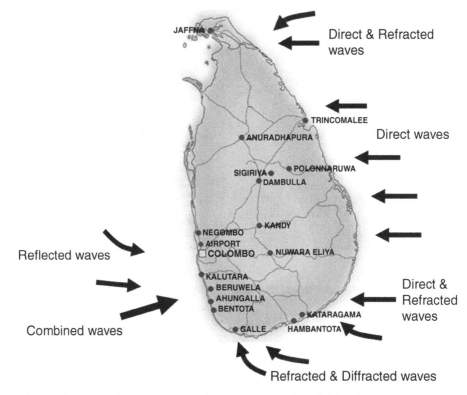

Figure 16.2 Coastal processes around Sri Lanka (Hettiarachchi and Samarawickrama, 2005).

that was transmitted over the shelf came directly towards land, as the Sri Lankan continental shelf is not wide enough to contribute towards significant energy dissipation. Discontinuities in the shelf would have caused problems as witnessed at the southern tip of the country. Waves diffracting around the southern parts of the island were further transformed by the complex wave patterns arising from these discontinuities, leading to greater impacts (Figure 16.1b). These near-shore transformation processes and shoreline geometry (e.g. bays, inlets and headlands) increased the wave heights along many parts of the southern and western provinces, which would normally have received only diffracted waves. Tsunami waves reflected off the Maldivian atolls also contributed to the complexity of the wave patterns. The effects of the tsunami were mitigated by coastal features such as sand dunes and vegetation but intensified by gaps created in coral reefs through illegal mining. The essentially flat inland topography and deficiencies in drainage facilities further exacerbated the problem.

Figure 16.2 illustrates the typical transformation processes that were effective around the island. Maximum wave heights were around 3 to 7 m on the east coast, 3 to 11 m on the south coast and 1.5 to 6 m on the west coast, with corresponding

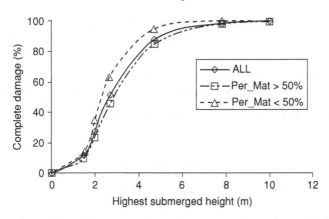

Figure 16.3 Vulnerability curves for various categories (Dias et al., 2009).

maximum penetration distances of around 2.5 km, 3 km and 1.8 km respectively (Wijetunga, 2006).

16.2 The Impacts

The primary loss of life was through drowning, while crushing injuries resulting from collapsing structures added to the casualties. At relatively low inundation depths of up to around 2 m, it was mainly 1 to 2 m-high boundary walls that succumbed. As depths increased, single-storey masonry structures, used for most of the housing stock along the coastline, were significantly damaged; and at inundation depths of around 4 m, they were completely swept off their foundations. The vulnerability curves in Figure 16.3, based on post-tsunami surveys (DCS, 2007), give the percentage of complete damage that can be expected for various wave heights. These percentages increase rapidly in the range between 2 and 4 m. Areas that had more than 50 per cent of houses with permanent materials (Per Mat > 50 per cent) had lower percentages of complete damage than areas with less than 50 per cent of houses with permanent materials (Per Mat < 50 per cent). Higher inundation depths of 3 to 5 m were accompanied by significant scouring, and a few two-storey structures failed through undermining of foundations. It should be noted that the main threats from a tsunami wave on a structure are overturning, sliding and scouring (Dias et al., 2006). Multi-storey buildings with reinforced concrete frames such as tourist hotels, government buildings and school buildings remained essentially intact.

On the one hand, the unforeseen tsunami affected all segments of society, including expatriate tourists on the coastline. On the other hand, it also intensified existing differences in social vulnerabilities. For example, it was the housing of the very poor that was most affected, both because of inferior quality and also because it was located virtually on the coast where land is not regulated (Khazai et al., 2006). Mortality rates among women were much greater than among men (Oxfam, 2005) because of a relatively lower ability to swim or climb trees – abilities seen as 'unsuitable' for women.

The temples, churches and mosques of all four major Sri Lankan religions serve as social focal points in everyday life. During the tsunami, they operated as locations of refuge because of their fairly robust construction and offered good protection.

16.3 The Aftermath

The media showed commendable initiative immediately after the event, requesting the public to donate food, bottled water and clothing for tsunami victims to designated collection centres, and this resulted in the distribution of spontaneous and effect-ive emergency supplies. Debris management was not satisfactory and caused much pollution of inland water bodies through indiscriminate dumping. Some foresight could have resulted in it being recycled, for example as crushed fine aggregate for new construction, especially given the pressure on raw material supply for the post-tsunami construction of around 100,000 houses (Khazai et al., 2006). Contamination of wells by saline water also took a long time to reduce. An environmental assessment was completed quite quickly, and recommendations were made to recover from the adversity (UNEP, 2005).

The temporary housing that was initially constructed was inadequate with respect to both privacy and comfort. The erection of permanent housing took around three years or more to complete. Construction was completed at different rates in individual districts. This could be correlated to many factors, the main one being availability of land (since reconstruction was not permitted on the coast). Other relevant factors were identified as the availability of human resources (e.g. skilled labour), political and administrative leadership, raw materials for construction, infrastructure and fin-ance. Table 16.1 has been developed based on responses from officials involved in reconstruction and gives the importance of these factors and their availability in three districts in Sri Lanka (Dias and Majeed, 2009). The new housing is of variable quality but in general more robust that the previously destroyed stock. Some reconstruction incorporated specific features to counter tsunamis. For example, Figure 16.4a shows a single-storey masonry chalet that was completely swept away by the tsunami; when a newly constructed chalet was built on the same site (Figure 16.4b), it was elevated in order to allow any future tsunami waters to flow under it. Some police stations and administrative offices along the south-west coast have also been reconstructed in similar fashion, with the open ground level used for parking.

Transport links were restored fairly quickly, the rail network being a particular case in point – a length of 120 km on the south-west coast was restored within two months (Ratnasooriya et al., 2007). Community-based warning systems are also in place now, and two successful evacuations to high ground have taken place after

Table 16.1. *Factors influencing rate of construction and weighted scores for districts related to percentage construction one year after the tsunami (modified from Dias and Majeed, 2009)*

Factor	Average weight	Factor availability for various districts		
		Ampara	Galle	Hambantota
Land	0.400	1.20	2.50	4.40
Human resources	0.145	2.40	2.67	3.40
Leadership	0.145	1.60	3.00	5.00
Raw materials	0.120	2.20	3.17	4.00
Infrastructure	0.100	1.80	3.00	2.60
Finance	0.090	4.40	4.50	4.60
Total Weighted Score		**1.900**	**2.908**	**4.132**
Houses Constructed/Assigned		**17.4%**	**41.0%**	**66.0%**

tsunami warnings following high-intensity earthquakes in the Sunda trench in March 2005 and September 2007.

Sri Lankan academics, researchers and professionals were able to collaborate with the many groups of international experts that conducted post-tsunami observations and surveys in the country, resulting both in knowledge generation and technology transfer. They have also been contributing their own expertise globally, for example, on developing the Indian Ocean Tsunami Warning System coordinated by the UNESCO Intergovernmental Oceanographic Commission in Paris. Professional institutions conducted many seminars regarding guidelines for reconstruction of buildings prone to natural disasters, and the first printed guideline was produced ten months after the

(a)   (b)

Figure 16.4 (a) Single-storey chalet completely swept away by the tsunami and (b) newly constructed elevated beach chalet on the identical site (photos by W. P. S. Dias).

event by the Society of Structural Engineers (2005). A reconstituted Disaster Management Centre (DMC) was commissioned to cover not only tsunami hazards but also a range of others.

The extent to which the various guidelines were implemented varied considerably, as there were many agencies involved in reconstruction and no mechanisms for strict enforcement. This is an issue that is still being tackled by the DMC, especially because quality and safety enhancements are difficult to incorporate in low-cost housing. 'No build' zones ranging from 100 m to 200 m from the coastline were initially declared but subsequently relaxed to the setback distances previously specified by the Coast Conservation Department, based largely on coastal erosion phenomena. Various research studies, both experimental and numerical, were carried out on the effectiveness of coastal vegetation as a tsunami barrier (Tanaka et al., 2009). Although some vegetation has been put in place, there is scope for more, because such bioshields serve multiple purposes and are also very cost effective.

Risk assessment is another area that received attention, a detailed analysis being carried out for the badly affected southern city of Galle, encompassing both physical and social vulnerability (Villagran, 2008). International collaboration has led to much documentation, for example under UNESCO auspices (e.g. UNESCO IOC, 2009a, 2009b) and also resulted in many conferences and workshops for training in disaster risk reduction, relating not merely to tsunamis but to a variety of other hazards, too. Hence, although the tsunami was disastrous in its consequences, it has hopefully contributed to greater public and institutional awareness and preparedness for the future.

16.4 The Lessons

One of the lessons learned was that dealing with natural disasters has to be a multi-disciplinary effort, requiring inputs from engineers, scientists, economists, sociologists, administrators, medical personnel and others. In ocean-wide phenomena such as tsunamis, regional and international co-operation is required, too. Furthermore, the phenomena require sophisticated technical analyses, but the results thereof must be transmitted to grassroots communities that are the most vulnerable.

Another lesson is that countering disasters involves multi-pronged approaches. For example, early warning systems and evacuation procedures are required primarily to protect life, but reducing physical vulnerability is also needed to minimise economic losses. Here too, 'hard' solutions such as coastal structures and strengthened buildings can be combined with 'soft' approaches such as coastal vegetation.

Finally, it has to be acknowledged that completely unexpected phenomena cannot be planned for. Nevertheless, one way that vulnerability to such events can be reduced is if assets and infrastructure are geographically distributed rather than concentrated. In this context, the construction of a highway linking the south of the country to the capital city of Colombo is a welcome development (although planned well before the tsunami), because its trace is well away from the coast, along which most of the current transport and economic infrastructure is situated. On the other hand, the city of Colombo itself is fast becoming a mega-city, with no other city elsewhere in the country being of any comparable size. This may be increasing the vulnerability of the country as a whole because of the possibility of an unforeseen disaster crippling its only major city.

References

ADB – Asian Development Bank (2005). *Sri Lanka 2005 Post-Tsunami Recovery Program, Preliminary Damage and Needs Assessment.* Colombo, Sri Lanka: Asian Development Bank, Japan Bank for International Cooperation, and World Bank. Accessed 25 October 2012 from: <http://www2.adb.org/Documents/Reports/Tsunami/ sri-lanka-tsunami-assessment.pdf>.

DCS – Department of Census and Statistics (2007). *Reports on Census of Buildings and People Affected by the Tsunami – 2004.* Colombo, Sri Lanka: Department of Census and Statistics. Accessed 25 October 2012 from: <http://www.statistics.gov.lk/tsunami/>.

Dias, P., Dissanayake, R. and Chandratilake, R. (2006). Lessons learned from tsunami damage in Sri Lanka. *Proceedings of ICE Civil Engineering*, 159, 74–81.

Dias, W. P. S. and Majeed, M. R. (2009). Factors influencing the rate of reconstruction after a tsunami disaster. Presentation at *International Conference on Accommodating Natural Hazards and Disasters in Social and Economic Infrastructure Development*, International Institute for Infrastructure Renewal and Reconstruction, Ahungalla, Sri Lanka, July.

Dias, W. P. S., Yapa, H. D. and Peiris, L. M. N. (2009). Tsunami vulnerability functions from field surveys and Monte Carlo simulation. *Civil Engineering and Environmental Systems*, 26, 181–194.

Hettiarachchi, S. S. L. and Samarawickrama, S. P. (2005). Experience of the Indian Ocean Tsunami on the Sri Lankan coast. In *Coastlines, Structures and Breakwaters 2005*, ed. N. W. H. Allsop. London, UK: Thomas Telford Publishing, pp. 19–27.

Khazai, B., Franco, G., Ingram, J. C. et al. (2006). Post-December 2004 tsunami in Sri Lanka and its potential impacts on future vulnerability. *Earthquake Spectra*, 22, S829-S824.

Oxfam (2005). *The Tsunami's Impact on Women. Oxfam Briefing Note.* London, UK: Oxfam International. Accessed 25 October 2012 from: <http://www.oxfam.org/sites/www. oxfam.org/files/women.pdf>.

Ratnasooriya, H. A. R., Samarawickrama, S. P. and Imamura, F. (2007). Post tsunami recovery process in Sri Lanka. *Journal of Natural Disaster Science*, 29, 21–28.

Society of Structural Engineers (2005). *Guidelines for Buildings at Risk from Natural Disasters.* Colombo, Sri Lanka: Society of Structural Engineers.

Tanaka, N., Nadasena, N. A. K., Jinadasa, K. B. S. N. et al. (2009). Developing effective vegetation bioshield for tsunami protection. *Civil Engineering and Environmental Systems*, 26, 163–180.

UNEP – United Nations Environmental Programme (2005). *Sri Lanka: Post-Tsunami Environmental Assessment.* Geneva: United Nations Environmental Programme.

UNESCO IOC – United Nations Educational, Scientific and Cultural Organisation, Intergovernmental Oceanographic Commission (2009a). *Hazard Awareness and Risk Mitigation in Integrated Coastal Management. IOC Manuals and Guides, No 50.* Paris: UNESCO.

UNESCO IOC – United Nations Educational, Scientific and Cultural Organisation, Intergovernmental Oceanographic Commission (2009b). *Tsunami Risk Assessment and Mitigation for the Indian Ocean. IOC Manuals and Guides, No 52*. Paris: UNESCO.

Villagran, J. C. (2008). *Rapid Assessment of Potential Impacts of a Tsunami: Lessons from the Port of Galle in Sri Lanka*. Bonn: UNU Institute for Environment and Human Security.

Wijetunga, J. J. (2006). Tsunami on 26 December 2004: spatial distribution of tsunami height and the extent of inundation in Sri Lanka. *Science of Tsunami Hazards*, 24, 225–239.

17

Recovery Efforts: The Case of the 2007 Cyclone Sidr in Bangladesh

BIMAL K. PAUL AND MUNSHI K. RAHMAN

On the night of 15 November 2007, Bangladesh was devastated by tropical cyclone Sidr. The cyclone swept across the south-western coast and ripped through the heart of the country from south-west to north-east with 155 m/h (248 km/h) winds, triggering up to 20 feet (6 m) tidal surges (Paul, 2009). Cyclone Sidr originated from a tropical depression in the Bay of Bengal on 11 November 2007 and quickly strengthened, reaching peak sustained winds of 135 m/h (215 km/h). This storm eventually hit offshore islands and made landfall on the south-eastern part of the Sundarbans Forest, and almost one-third of this forest, a world natural heritage site, was totally destroyed. After landfall, Sidr weakened into a tropical storm and dissipated on 16 November 2007 (GoB, 2008a).

Cyclone Sidr was one of the ten most devastating cyclones that struck Bangladesh during the 131 years between 1876 and 2007. According to the latest official reports, more than 27 million people in thirty districts suffered damage caused by Sidr, including damage to buildings, roads, bridges, culverts, ferries, embankments and crops (Paul and Dutt, 2010). Of the affected districts,[1] twelve coastal districts were hardest hit by this storm (Figure 17.1). Cyclone Sidr claimed 3,406 lives – strikingly fewer than anticipated, and some 55,000 people sustained physical injuries (GoB, 2008a). An estimated 1.87 million livestock and poultry perished, and crops on 2.4 million acres of land were completely or partially destroyed. In addition, nearly 1.5 million homes and 5,500 miles (8,075 km) of roads were badly damaged or destroyed (GoB, 2008a). The storm also caused power outages that resulted in a near-countrywide blackout lasting more than thirty-six hours (Natural Hazards Center, 2008).

The Joint Damage Loss and Needs Assessment mission, led by the World Bank, estimated the total cost of all damage caused by cyclone Sidr at US$1.7 billion, a figure that represents about 3 per cent of the total gross national product of Bangladesh (GoB, 2008a). More than two-thirds of the disaster effects were physical damages and

---

[1] A district is the second largest administrative unit in Bangladesh, with an average population of 2.5 million.

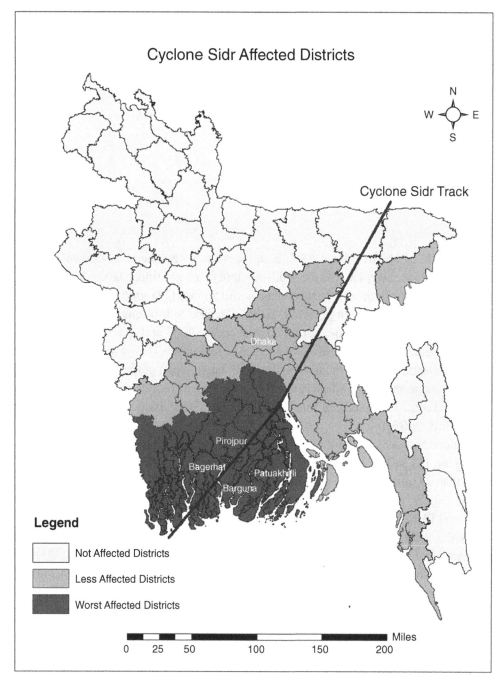

Figure 17.1 Cyclone Sidr affected districts (based on GoB 2008a, 2008b).

one-third were economic losses, with most damages and losses incurred in the private sector. Nearly two million people lost income and employment in the more severely impacted districts. The effects of the cyclone were highly concentrated by district. The Bangladesh Government identified Bagerhat, Barguna, Patuakhali and Pirojpur as the districts most severely affected by cyclone Sidr (Figure 17.1). All affected coastal districts also suffer from higher poverty rates than the national average (GoB, 2008a).

Five days before Sidr made landfall, the Bangladesh Government began to broadcast warnings on radio and television, and it issued emergency evacuation orders almost twenty-seven hours before landfall (Paul and Dutt, 2010). Unfortunately, only about 33 per cent of residents in the impacted coastal areas sought shelter at some location other than their own homes after receiving evacuation orders (GoB, 2008a). Sixty per cent of all evacuees took refuge in public cyclone shelters; the remaining 40 per cent took shelter either in public buildings (e.g. government offices, mosques and college) or in neighbours' houses that they perceived as being structurally stronger than their own.

The objective of this study is to explore the progress made to date toward recovery from the impacts of cyclone Sidr. More specifically, the emphasis of this study is on housing recovery because the impact of the cyclone on this sector has been extreme. Total damage to the housing sector is estimated at US$839 million, a figure representing more than 50 per cent of the total losses of all sectors (GoB, 2008a).

17.1 Sidr Recovery Efforts

Following cyclone Sidr, the Bangladesh Government initiated an early recovery program to meet the immediate needs of affected households and create a solid foundation for long-term recovery and the reconstruction phase. This early phase included interventions designed to provide social protection, infrastructure repair and income recovery. For efficient and effective recovery, the Bangladesh Government formed the Early Recovery Cluster Coordination Group with twenty United Nations organisations, representatives of government ministries and international and national non-governmental organisations (NGOs). The United Nations Development Program (UNDP) was appointed lead agency for the group, which worked closely with the Bangladesh Government's Ministry of Food and Disaster Management as co-chair (Lattimer and Pujiono, 2008).

Using a 'build back better' approach, the long-term recovery program of the Bangladesh Government consisted of two components: Reconstruction and upgrading of damaged social and economic infrastructure and recovery and improvement of agricultural and non-farm activities (GoB, 2008a). The Government decided to construct housing for 20,000 households, upgrading the transport network and rebuilding damaged or destroyed marketplaces, as well as reconstructing water supply and electric services. Other program goals included the reconstruction of urban public infrastructure, upgrading of health service infrastructure, reconstruction of fully damaged schools to school-cum-shelter, reconstruction and upgrading of damaged embankments and rehabilitation of the Sundarbans mangrove forest (GoB, 2008a).

Based on a careful review of materials published in print media, scholarly journals and Internet websites regarding recovery from the impacts of cyclone Sidr, it is evident that full recovery has yet to be achieved. In other words, affected households have not been able to reconstitute their assets and 'normalise' their livelihoods to pre–Sidr levels (also see TANGO International, 2010). Although the Bangladesh Government has achieved success in short-term recovery efforts (e.g. providing food security and transitional shelter assistance, ensuring supply of drinking water and controlling infectious diseases), it has thus far failed to achieve many of its long-term recovery goals. For example, all damaged roads and embankments have not been repaired, and many households are still living in temporary makeshift shelters.

17.2 Sidr Housing Recovery Efforts

Immediately after cyclone Sidr's landfall, the Government of Bangladesh distributed 13,406 tents and 5,000 pieces of plastic sheet to provide transitional shelters for those in need of shelter (GoB, 2008b).[2] Several NGOs and donor agencies also participated in recovery efforts by providing transitional shelters for Sidr survivors. For example, Muslim Aid has constructed 800 transitional shelters to date and provided 2,000 families with construction materials to repair their homes in Bagerhat, Patuakhali and Pirojpur districts in conjunction with the European Commission for Humanitarian Aid (ECHO) (Muslim Aid, 2007). Also, Habitat for Humanity Bangladesh has constructed 480 transitional shelters with proper sanitation using contributions from UNICEF Bangladesh and Habitat Great Britain. Many Sidr victims whose homes were destroyed or damaged also built transitional/temporary shelters entirely at their own expense (IFRCCS, 2008).

Along with distributing materials for temporary shelters, the Bangladesh Government made available a one-time housing grant of US$70 to some 100,000 families in the most severely impacted areas whose homes were completely destroyed. The government also distributed 13,000 bundles of corrugated galvanised iron sheets to Sidr survivors (GoB, 2008b). Although these items helped the most vulnerable families meet some urgent needs, the amount was not only insufficient to rebuild safe and strong homes, but the distribution of these items was also not always equitable (Chughtai, 2008). Moreover, many individuals did not use the grants for their original

---

[2] Following the landfall of cyclone Sidr, the Government of Bangladesh estimated that about one-third of those who lost their homes because of the cyclone need assistance in rebuilding their homes, while the remaining two-thirds need assistance in repairing their partially damaged homes (GoB, 2008a).

intended purpose; rather, they spent this grant money on emergency items such as food or winter clothing (GoB, 2008a).

In addition to providing transitional shelters and shelter repair assistance, the Bangladesh Government also decided to build cyclone-resilient core shelters for 78,519 households (Chughtai, 2008). This program was, in fact, the key component of the government's 'build back better' recovery plan. It introduced the Philippines-style Core Shelter Program to Sidr-impacted areas to improve construction quality of damaged houses. The basic component of this shelter is the construction of a small house of strong cyclone-resistant materials to which storage spaces, verandas and extra rooms can be added in a modular fashion over time. Consideration is being given to the height of the veranda and the possible provision of water-resistant storage containers for seeds, grains and items of value, such as identity and loan documents (GoB, 2008a). Unfortunately, the construction of these core shelters has been slow, and by the middle of November 2008, just more than half the planned shelters had been completed (Chughtai, 2008).

Similar to the Bangladesh Government, UNDP also committed to building 15,600 core homes in Sidr-impacted areas. As of the middle of August 2010, UNDP had built more than 9,000 cyclone-resilient core homes in the most severely impacted districts. Designed according to local environmental and cultural needs, these new structures allow for future expansion, giving households options for investing their resources into expanding their homes. These shelters also include gutters for harvesting rainwater and are constructed to withstand a Category 4 cyclone (UNDP, 2010).

Several NGOs (e.g. Concern Worldwide, Habitat for Humanity, Muslim Aid, World Vision and the Bangladesh Rural Advancement Committee – BRAC) and foreign countries, such as India, Kuwait and Saudi Arabia, have also promised to build approximately 82,000 core shelters for families affected by Sidr (Heinrich, 2008). Saudi Arabia alone has agreed to fund construction of about 21,000 houses. Muslim Aid has already provided 1,000 permanent core shelters in Bagerhat and Patuakhali districts. BRAC, with help from Oxfam, built sturdy homes for 400 families who survived cyclone Sidr (GoB, 2008b).

In looking for cost-effective solutions to increase the safety of housing for poorer families, several NGOs are focusing their support on households wishing to strengthen plinth construction and to elevate their plinths for protection against storm surge – for example, putting dwellings on strong stilts. Additionally, there are plans to implement 'cash for training' programs to educate carpenters, masons and households in the use of cyclone-resistant construction techniques, including simple forms of wind bracing (GoB, 2008a). One local NGO has developed a design for a two-storey house on stilts, with an upper floor veranda on which livestock may be sheltered during floods and cyclones (Davidson, 2008).

Unfortunately, even four years after cyclone Sidr, nearly one million people in Sidr-affected areas have been living in makeshift shelters built from a variety of available materials including banana leaves, cloth and salvaged wood and iron. To date, communications with officials from the Ministry of Food and Disaster Management of the Bangladesh Government reveal donor agencies, aid organisations and government agencies have only built some three-quarters of the permanent housing units they initially promised. UNDP still needs to construct 6,600 additional permanent homes for the most vulnerable families in the most severely impacted districts (UNDP, 2010). Lack of funding is a serious challenge for the Bangladesh Government for achieving its goal of building 78,519 core shelters. While many volunteer and other organisations have already built and/or have a plan to build secure homes for coastal residents, no one of these organisations is responsible for building core shelters.

For national and foreign NGOs and other organisations, transportation and communication have been tremendous obstacles to building the promised number of permanent houses for cyclone Sidr survivors. A single core shelter requires 4,500 bricks, up to twenty bags of cement and a roof truss that weighs almost 16 tons (3,520 lbs.). Compiling these materials and bringing them to construction sites in remote and isolated locations has proved a challenging task. Transporting these materials is not only time consuming but costly, too. Additionally, in some locations, there have been severe shortages of masons, carpenters and welders. All these factors have been instrumental in delaying the construction of these new, cyclone-resistant core homes (UNDP, 2010).

17.3 Conclusions

Based on information collected from secondary sources, this chapter has examined the progress thus far made toward housing recovery by cyclone Sidr survivors in coastal Bangladesh. A careful review of published materials and conversations with officials of the Bangladesh Government and relevant NGOs suggest that full housing recovery clearly has yet to be achieved. Progress towards full recovery has been slow primarily because of funding constraints and lack of roads to transport building materials. Fortunately, the government has been trying to convince donor agencies to provide funding committed after cyclone Sidr's landfall and improve the road network in the coastal region. Residents of Sidr-affected areas who are now living in houses built by both private and public organisations feel that they are better protected against future cyclones. Other residents of the area deserve and expect similar protection before a cyclone strikes again.

References

Chughtai, S. (2008). *One Year After Cyclone Sidr: Fear Replaces Hope. Oxfam Briefing Note*. London, UK: Oxfam International. Accessed 26 October 2012 from: <http://policy-practice.oxfam.org.uk/publications/one-year-after-cyclone-sidr-fear-replaces-hope-114558>.

Davidson, S. (2008). *A Review of the IFRC-led Shelter Coordination Group: Bangladesh Cyclone Sidr Response 2007–2008*. Dhaka: Bangladesh Red Crescent Society.

GoB – Government of Bangladesh (2008a). *Cyclone Sidr in Bangladesh: Damage, Loss and Needs Assessment for Disaster Recovery and Reconstruction*. Dhaka: Government of Bangladesh.

GoB – Government of Bangladesh (2008b). *Super Cyclone Sidr 2007: Impacts and Strategies for Interventions*. Dhaka: Ministry of Food and Disaster Management.

Heinrich, C. (2008). *Sidr Survivors Begin to Rebuild Their Lives*. Boston: Oxfam America. Accessed 26 October 2012 from: <http://www.oxfamamerica.org/articles/sidr-survivors-begin-to-rebuild-their-lives>.

IFRCCS – International Federation of Red Cross and Red Crescent Societies (2008). *Cyclone Sidr in Bangladesh: Shelter Plan of Action*. Dhaka: International Federation of Red Cross and Red Crescent Societies.

Lattimer, C. and Pujiono, P. (2008). Early recovery in Bangladesh: linking response and early recovery. *UNDP Newsletter*, 1–2.

Muslim Aid (2007). *After Cyclone Sidr*. Dhaka: Muslim Aid.

Natural Hazards Center (2008). Cyclone Sidr – Bangladesh. *Natural Hazards Observer* (3), 4.

Paul, B.K. (2009). Why relatively fewer people died? The case of Bangladesh's Cyclone Sidr. *Natural Hazards* 50(2), 289–304.

Paul, B.K. and Dutt, S. (2010). Hazard warnings and responses to evacuation orders: the case of Bangladesh's Cyclone Sidr. *Geographical Review* 100(3), 336–355.

TANGO – Technical Assistance to Non-Governmental Organizations International (2010). *An Assessment of Livelihood Recovery. Report Presented to: Save the Children Bangladesh*. Tucson, AZ: Save the Children-USA and Partners.

UNDP – United Nations Development Programme (2010). *New Homes for Cyclone*. Dhaka: United Nations Development Programme.

18

Coffee, Disasters and Social-Ecological Resilience in Guatemala and Chiapas, Mexico

HALLIE EAKIN, HELDA MORALES, EDWIN CASTELLANOS, GUSTAVO CRUZ-BELLO, AND JUAN F. BARRERA

The Mesoamerican region is particularly susceptible to hazards and natural disasters (Dilley et al., 2005). In early October 2005, Stan developed as the twentieth cyclone of a record-breaking hurricane season in the Atlantic (Unisys Weather Information System, 2010), causing great devastation and human loss. In Guatemala, an estimated half a million people were directly affected and total economic losses were calculated at US$983 million, equivalent to 3.4 per cent of the GDP for 2004 (CEPAL, 2005). Over the last decade, the region has been hit by a series of substantial storms. In 1998, hurricane Mitch poured heavy rains on saturated soils and produced a toll of 268 dead and 121 missing persons and economic losses of US$748 million in Guatemala alone (CEPAL, 1999; 2010). In May of 2010, tropical storm Agatha hit southern Mexico and Guatemala, once again causing a series of mudslides and flooding that resulted in ninety-six casualties and sixty-two missing persons. Climate change scenarios for the region suggest damages will continue as global temperatures rise (Webster et al., 2005; Knutson et al., 2010), and the threat of continued losses is increasingly viewed as a threat to national security for the countries of the region (Fetzek, 2009).

The determinants of vulnerability in the region are complex. Despite the level of losses associated with mudslides and flooding, hurricane landfalls are surprisingly infrequent in Guatemala and the southern Mexican state of Chiapas (Jáuregui, 2003). Hurricane winds are typically slowed to the level of tropical storms or depressions by the mountainous terrain of the region. This terrain, with sandy, volcanic soils, is nevertheless highly unstable, resulting in mudslides that are the main cause of the casualties and destruction (Restrepo and Alvarez, 2006). The populations that make their livelihoods on these steep slopes are often the most sensitive to these impacts, yet as resource managers with long histories in the region, they are also a potentially vital source of social and ecological resilience. In this case study, we explore the opportunities for enhancing disaster preparedness through better use of local knowledge and the social infrastructure of some of the regions' more vulnerable populations: The smallholder coffee farmers that make their livings on land highly susceptible to mudslides, erosion and flooding.

18.1 Coffee and Disaster

The majority of coffee producers in Mesoamerica have less than 5 ha of land, are typically asset poor and often live in relatively isolated communities with constrained access to public services and markets (Goodman, 2008; Tucker et al., 2009). During the harvest period, many households depend on migrant labour. Producers rely on intermediaries, who travel to their villages over difficult terrain, to purchase their harvests. Their vulnerability is thus not only a product of their physical exposure and reliance on relatively sensitive natural resources but also their livelihood dependence on the movement of labour, goods and services through infrastructure and transport networks. These transport networks are highly susceptible to damage and disruption from hazards, exacerbating losses and making rescue and recovery efforts more difficult.

Widespread losses in coffee as a direct or indirect result of hazards often translate into broader-scale economic crises given the importance of coffee to agricultural exports, employment and economic productivity in the regional economy (CEPAL, 2002). In Chiapas alone, an estimated 75,000 ha of coffee were directly damaged by Stan, representing more than a third of Chiapas's planted area in coffee (CENAPRED, 2006). Many more households were further affected by lack of access to transport, health facilities and food.

Yet while the coffee sector is highly sensitive to loss, it may also be a source of social and ecological resilience (Eakin et al., 2006). Among other characteristics, resilient systems tend to support high diversity and capacity for self-organisation and learning and to feature an active flow of resources, information and knowledge across spatial scales and organisational levels (Folke, 2006). Farmers have benefited from participating in coffee markets, particularly for farmers participating in niche commodity chains (Bacon et al., 2008). Shade-grown coffee plantations are associated with providing habitat conditions for a diversity of flora and fauna (Perfecto et al., 1996; Greenberg et al., 1997; Soto-Pinto et al., 2001; Williams-Guillén et al., 2006). Coffee plantations are also considered sources of ecological services including pest management, pollination, rainfall capture, erosion control and carbon sequestration (De Jong et al., 1995; Perfecto et al., 2004; Philpott et al., 2008a, 2008b; Jha and Vandermeer, 2009). Some of these services may be an advantage in the face of meteorological extremes (Holt-Giménez, 2002; Lin, 2007). For example, Holt-Giménez (2002) found that plots managed on agroecological principles suffered less erosion in face of heavy rainfall events, while Lin (2007) found that coffee cultivated with agroforestry principles effectively mitigated against microclimate and soil moisture variability.

The farmers themselves are also a resource. While significantly constrained by poverty and often preoccupied with family health issues, debt and meeting daily subsistence requirements, many coffee farmers have insights into the biophysical drivers of their vulnerability and the type of land use and livelihood strategies that might enhance their resilience. For example, in a case study of farmers' losses and responses following Stan in Chiapas (Cruz-Bello et al., 2011), approximately a third of respondents in the surveyed communities attributed the disaster to factors including pollution, deforestation and lack of land management. This study also demonstrated that household members tended to believe land parcels that were relatively homogenous, un-terraced and without shade trees would be more affected than a more complex vegetated landscape. This confirms the findings of scholars such as Philpott et al. (2008a), who found that landscapes supporting more complex vegetation were less subject to landslides, and Holt-Giménez, (2002) and Lin (2007), who found that terracing and shade cover were effective in mediating moderate climatic variability.

There is also a growing body of work that suggests that social organisation can be both directly and indirectly instrumental in building adaptive capacity (e.g. Pelling and High, 2005). An increasing number of farmers in the region are now organising in cooperatives as a means of managing market risk, and there are some examples of cooperatives organising to address climate change. In our own research, we have found that cooperatives are valued by farmers because of the improved access to information offered by these organisations; cooperatives also offer the institutional structure for communication and social mobilisation to alter livelihoods and land use.

The group Más Café in Chiapas, for example, has the support of the German bilateral development agency GTZ (Deutsche Gesellschaft für Technische Zusammenarbeit) and Cafédirect, a British company specialising in fair trade, to implement climate adaptation actions (Más Café, 2010). A pilot project was developed with two communities belonging to Más Café that included five activities as part of an adaptation strategy: (1) maintain and increase the forest cover around the communities; (2) improve the pest management plans; (3) maintain and increase the amount of carbon sequestration in coffee plantations; (4) promote the use of renewable energies in the processing of harvested beans; and (5) improve the drying process for coffee. In Guatemala, coffee organisations such as UPAVIM (which stands for 'united to live better') in Atitlán are working together with local and international NGOs to develop adaptation strategies that include disaster management procedures for extreme weather events. UPAVIM is also working to reduce the vulnerability of associated farmers through strategies to increase and diversify sources of income. They have successfully certified their production as organic and are seeking certification with Rainforest Alliance in their newly developed 'climate friendly' seal (Sustainable Agriculture Network, 2011) with the hope of improving the price for their coffee. They have also diversified their income by developing an eco-tourism activity offering a coffee tour of their plantations.

Belonging to a cooperative typically entails engaging with a specific niche market for coffee (e.g. organic, fair trade, bird-friendly) and altering land use management practices to comply with the demands of those specific markets. Land terracing, increasing shade density and variety and building up organic matter through management of understory vegetation are all practices that are often encouraged. Again, scholars have found some evidence that some of these practices enhance the resilience of landscapes in the face of a range of disturbances and thus may well reduce livelihood vulnerability to many cyclone events (Lin, 2007).

18.2 Disaster Management and Response

Despite the increasing scientific attention to the diversity of ecological services coffee farming might offer, as of yet there is little evidence these services are being considered in disaster risk management planning and policy. Disaster management in the region has primarily focused on enhancing the efficiency of early warning systems and emergency response (Saldaña-Zorrilla, 2008). The public sector efforts in both Guatemala and Chiapas following Stan, for example, focused on improving emergency response in order to avoid loss of life and meet basic needs (e.g. Gobierno de Chiapas, 2010). Where attention has been paid to agricultural rehabilitation, the emphasis has primarily been on reconstruction of damaged infrastructure (e.g. roads, wet mills, patios to dry beans) and the immediate needs of farmers (e.g. credit, cash for labour) (SAGARPA, 2005) and, to a lesser extent, watershed reforestation.

The recent impact of tropical storm Agatha underscores the difficulty in reducing vulnerability by focusing primarily on emergency response. Infrastructure, water supply, and coffee plantations were once again severely damaged. In Chiapas, the public sector was criticised for its apparent delayed response to the needs of some affected rural communities and for the lack of maintenance of the drainage infrastructure. While no lives were lost, the extensive damage experienced suggests that the lessons from disaster impact at both the community and government level can be difficult to maintain in the forefront of day-to-day operations. In Guatemala, Agatha's impact was all the worse because the reconstruction from Stan's damage was still incomplete.

In short, while there have been efforts to improve disaster response over the last decade, there is little evidence in the region that attention has been paid to empowering local communities to act on their existing knowledge and experience in relation to risk or to building on local social institutions to enhance local risk management capacity. In fact, some research suggests that authorities tend to view local communities, their land use practices and attitudes as part of the problem rather than the solution (Saldaña-Zorrilla, 2008).

Part of this weakness in disaster response appears to be that disaster management is principally considered an issue of social protection and preparedness for specific extreme events, removed from the broader spatial and temporal context of social-ecological development. Altering conditions of chronic poverty and social marginalisation is a long-term process, and tangible results are often only evident over longer time frames. Nevertheless, in the coffee-producing areas of Mesoamerica, sources of resilience may well be found in farmers' existing ecological knowledge, their capacities as resource managers and their social institutions. Given the communication challenges and relative isolation of many communities that are often subject to extreme events, important gains in disaster preparedness might be made through public sector efforts that build on existing social capital, organisation and knowledge to enhance ecological services and human response capacity.

Acknowledgements

The work referred to in this case study was carried out in part with the aid of a grant from the Inter-American Institute for Global Change Research CRN-2060, which is supported by the US National Science Foundation (Grant GEO-0452325), and in part with support from a collaborative research grant from UC-MEXUS. The authors appreciate the additional support of three communities in Siltepec, Chiapas, who tolerated the research initiative even while recovering from Stan.

References

Bacon, C., Mendez, E., Gliessman, S., Goodman, D. and Fox, J. (2008). *Confronting the Coffee Crisis: Fair Trade, Sustainable Livelihoods and Ecosystems in Mexico and Central America.* Cambridge: Massachusetts Institute of Technology.

CENAPRED – Centro Nacional de Prevención de Desastres (2006). *Características e Impacto Socioeconómico de los Huracanes "Stan" y "Wilma" en la República Mexicana en el 2005.* Mexico City: Secretaría de Gobernación and Centro Nacional de Prevención de Desastres.

CEPAL – Comisión Económica para América Latina y el Caribe (1999). *Guatemala: Assessment of the Damage Caused by Hurricane Mitch, 1998.* Mexico City: Comisión Económica para América Latina y el Caribe.

CEPAL – Comisión Económica para América Latina y el Caribe (2002). *Centroamérica: El Impacto de la Caída de los Precios del Café en 2001.* Mexico City: Comisión Económica para América Latina y el Caribe.

CEPAL – Comisión Económica para América Latina y el Caribe (2005). *Efectos en Guatemala de las Lluvias Torrenciales y la Tormenta Tropical Stan, Octubre de 2005.* Mexico City: Comisión Económica para América Latina y el Caribe

CEPAL – Comisión Económica para América Latina y el Caribe (2010). *Guatemala Efectos del Cambio Climático sobre la Agricultura.* Mexico City: Comisión Económica para América Latina y el Caribe.

Cruz-Bello, G., Eakin H., Morales, H. and Barrera, J. F. (2011). Linking multi-temporal analysis and community consultation to evaluate the response to the impact of Hurricane Stan in coffee areas of Chiapas, Mexico. *Natural Hazards*, 58(1), 103–116.

De Jong, B., Montoya G., Nelson, K. et al. (1995). Community forest management and carbon sequestration: a feasibility study from Chiapas, Mexico. *Interciencia*, 20(6), 409–416.

Dilley, M., Chen, R. S., Deichmann, U. et al. (2005). *Natural Disaster Hotspots: A Global Risk Analysis.* Washington, DC: World Bank.

Eakin, H., Tucker C. and Castellanos E. (2006). Responding to the coffee crisis: a pilot study of farmers' adaptation in Mexico, Guatemala and Honduras. *The Geographical Journal*, 172(2), 156–171.

Fetzek, S. (2009). *Impactos Relacionados con el Clima en la Seguridad Nacional en México y Centroamérica*. London, UK: Royal United Services Institute.

Folke, C. (2006). Resilience: the emergence of a perspective for social-ecological systems analyses. *Global Environmental Change*, 16, 253–267.

Gobierno de Chiapas (2010). *Programa Emergente Huracán Stan, Cuenta Hacienda Pública Estatal*. Chiapas: Gobierno de Chiapas, Secretaria de Hacienda. Accessed 29 October 2012 from: <http://www.haciendachiapas.gob.mx/rendicion-ctas/cuentas-publicas/informacion/CP2005/TomoI/Stan.pdf>.

Goodman, D. (2008). The international coffee crisis: a review of the issues. In *Confronting the Coffee Crisis, Fair Trade, Sustainable Livelihoods and Ecosystems in Mexico and Central America*, eds. C. M. Bacon, V. E. Méndez, S. R. Gliessman, D. Goodman, and J. A. Fox. Cambridge: Massachusetts Institute of Technology, pp. 3–25.

Greenberg, R., Bichier P., Cruz, A. and Reitsma, R. (1997). Bird populations in shade and sun coffee plantations in Central Guatemala. *Conservation Biology*, 11(2), 448–459.

Holt-Giménez, E. (2002). Measuring farmers' agroecological resistance after Hurricane Mitch in Nicaragua: a case study in participatory, sustainable land management impact monitoring. *Agriculture, Ecosystems and Environment*, 93, 87–105.

Jáuregui, E. (2003). Climatology of landfalling hurricanes and tropical storms in Mexico. *Atmósfera*, 16, 193–204.

Jha, S. and Vandermeer, J. (2009). Contrasting bee foraging in response to resource scale and local habitat management. *Oikos*, 118, 1174–1180.

Knutson, T., McBride J. L., Chan J. et al. (2010). Tropical cyclones and climate change. *Nature Geoscience*, 3, 157–163.

Lin, B. (2007). Agroforestry management as an adaptive strategy against potential microclimatic extremes in coffee agriculture. *Agriculture and Forest Meteorology*, 144, 84–94.

Más Café (2010). *AdapCC – Adaptación al Cambio Climático Para Pequeños Productores de café y té*. Accessed 29 October 2012 from: <http://www.adapcc.org/index_es.htm>.

Pelling, M. and High, C. (2005). Understanding adaptation: what can social capital offer assessments of adaptive capacity? *Global Environmental Change Part A*, 15(4), 308–319.

Perfecto, I., Rice, R., Greenberg, R. and Van der Voort, M. (1996). Shade coffee: a disappearing refuge for biodiversity. *Bioscience*, 46(8), 598–608.

Perfecto, I., Vandermeer, J., Lopez Bautista, G. et al. (2004). Greater predation in shaded coffee farms: the role of resident neotropical birds. *Ecology*, 85, 2677–2681.

Philpott, S., Lin, B., Jha, S. and Brines, S. (2008a). A multi-scale assessment of hurricane impacts on agricultural landscapes based on land use and topographic features. *Agriculture, Ecosystems and Environment* 128, 12–20.

Philpott, S., Perfecto, I. and Vandermeer, J. (2008b). Behavioral diversity of predatory arboreal ants in coffee agroecosystems. *Environmental Entomology*, 37, 181–191.

Restrepo, C. and Alvarez, N. (2006). Landslides and their contribution to land-cover change in the mountains of Mexico and Central America. *Biotropica*, 38, 446–457.

Saldaña-Zorrilla, S. O. (2008). Stakeholders' views in reducing rural vulnerability to natural disasters in Southern Mexico: hazard exposure and coping and adaptive capacity. *Global Environmental Change*, 18, 583–597.

SAGARPA – Secretaria de Agricultura, Ganadería, Desarrollo Rural, Pesca y Alimentación (2005). *Prioriza SAGARPA Apoyos para Cafeticultores Afectados por Huracán Stan en Chiapas*. Boletín SAGARPA No. 322/055. Coordinación General de Comunicación Social, SAGARPA. México, D.F. Accessed 11 June 2013 from: <http://www.cedrssa.gob.mx/includes/asp/download.asp?iddocumento=383&idurl=202>.

Soto-Pinto, L., Romero-Alvarado, Y., Caballero-Nieto, J. and Segura, G. (2001). Woody plant diversity and structure of shade-grown-coffee plantations in Northern Chiapas, Mexico. *Revista de Biologia Tropical*, 49(3–4), 977–987.

Sustainable Agriculture Network (2011). *SAN Climate Module. Criteria for Mitigation and Adaptation to Climate Change*. San Jose, Costa Rica: Sustainable Agriculture Network. Accessed 29 October 2012 from: <http://sanstandards.org/sitio/subsections/display/51>.

Tucker, C., Eakin, H. and Castellanos, E. (2009). Perceptions of risk and adaptation: coffee producers, market shocks, and extreme weather in Central America and Mexico. *Global Environmental Change*, 20, 23–32.

Unisys Weather Information System (2010). *Hurricane/Tropical Data*. Malvern, PA: Unisys Weather Information System. Accessed 26 October 2012 from: <http://weather.unisys.com/hurricane/index.html>.

Webster, P. J., Holland, G. J., Curry, J. A. and Chang, H. R. (2005). Changes in Tropical cyclone number, duration, and intensity in a warming environment. *Science*, 309, 1844–1846.

Williams-Guillen, K., McCann, C., Martinez-Sanchez, J. and Koontz, F. (2006). Resource availability and habitat use by mantled howling monkeys in a Nicaraguan coffee plantation: can agroforest serve as core habitat for a forest mammal? *Animal Conservation*, 9(3), 331–338.

19

Responding to Floods in the Nile basin: A Case Study of the 1997–1998 Floods in the Upper White Nile

MARISA C. GOULDEN AND DECLAN CONWAY

19.1 Flood Response and Adaptation to Climate Change

Flood events are an important feature of Nile River hydrology and represent a serious challenge to water resource management in the basin. Improved understanding of the causes, consequences and responses to floods can support the planning of strategies to adapt to future climate change, which may cause potentially rapid but highly uncertain changes in flood frequency and intensity. Events such as the Mozambique floods in 2000 (Christie and Hanlon, 2001) and Sahel floods (Tarhule, 2005; Tschakert et al., 2010) highlight the significance of floods in Africa across a range of scales.

The Nile and its sub-basins have exhibited decadal scale variability in streamflow since modern instrumental records began in the late 1860s (Conway, 2005). These events have triggered technical and institutional responses, including establishment of the Nile Forecasting Department in Egypt to improve management of successive years of low flows (Conway, 2005). Basin scale floods, manifest as peak flow years in the Main Nile record, occurred in 1878, 1947 and 1961 (Conway, 2002). Smaller-scale events include flooding on the Blue Nile in 1988 (Sutcliffe et al., 1989) and the widespread floods in the White Nile system in 1997 and 1998 (Conway et al., 2005; Goulden, 2006) that form the basis of this study. Recent flood events have occurred in Ethiopia, Sudan and Uganda in 2006 and 2007 (WaterAid, 2008; Jury, 2010; Moges et al., 2010).

In this chapter, we examine the impacts of and responses to a major flood event in the Upper White Nile Basin during 1997/98. In the context of adaptation, study of past flood events can be used to understand patterns of vulnerability to flooding-induced disasters and modes of response as a surrogate for how such processes may play out with climate change in the future (Few, 2003; Penning-Rowsell et al., 2006) to build understanding of factors contributing to resilience to floods (López-Marrero and Tschakert, 2011) and make linkages between strategies for disaster risk reduction and climate change adaptation (Schipper and Pelling, 2006; Thomalla et al., 2006).

We use the term 'adaptation' to refer to a range of actions taken by people in response to or in anticipation of a change, stress or natural hazard in order to manage its impact and maintain well-being (Smit and Wandel, 2006; Goulden et al., 2009). Adaptation can contribute sources of resilience to a system, where 'resilience' refers to the ability of a system to absorb shocks whilst maintaining its functions or re-organise by drawing on buffer capacity, adaptive capacity and self-organisation (Walker et al., 2006; López-Marrero and Tschakert, 2011). We use the term 'vulnerability' to mean 'the state of susceptibility to harm from exposure associated with environmental and social change and from the absence of capacity to adapt' (Adger, 2006).

19.2 Flooding in the White Nile Around Lake Kyoga and Lake Victoria in 1997 and 1998: Impacts and Responses

Description of the Flood Event and Its Impacts

Extreme rainfall occurred between November 1997 and February 1998 over East Africa, causing a sharp rise in lake levels across the region (Birkett et al., 1999). Detailed analysis of rainfall records by Conway (2002) demonstrated that the event was most pronounced from 10 °N to 10 °S, particularly around 0 to 5 °S, where it was the wettest year on record (from 1900 to 2000), and 10 to 5 °N (third wettest on record). The most extreme rainfall occurred primarily during October and November 1997. Lake Victoria's level increased in response to the event by roughly 1 m between 1997 and 1998, the second largest annual increase since 1900 (the largest was 1.2 m 1961–62). Lake Kyoga in the White Nile system rose by 1.3 m between October 1997 and May 1998.

The heavy rains and associated flooding caused widespread disruption across East Africa (see Table 19.1 for examples). Flooding impacts were widespread around Lake Kyoga because of the low-lying landscape and a blockage at the outlet of the lake caused by floating vegetation dislodged by the floods. This blockage combined with the high inflow into the lake caused a maximum rise of 2.3 m from pre–October 1997 levels by June 1999 and a 10 per cent expansion in area of the lake and surrounding wetlands (Gumbricht, 2005; Goulden, 2006). Lake Wamala rose by 2.8 m and expanded to almost three times the previous area in the same time period[1] (Goulden, 2006). Similar-scale floods in the early 1960s were the most extreme on record. Both events have been linked to enhanced convection in the western Indian Ocean (Birkett et al., 1999; Webster et al., 1999). The impacts of the 1997/98 floods were arguably more severe in terms of socio-economic effects than those of the 1960s because of the large

---

[1] Satellite images and aerial photographs indicate that the lake area expanded from 74 km² in 1995 to 217 km² by 1999. Lake level data indicate that the majority of this expansion occurred between October 1997 and May 1998 (Goulden, 2006).

Table 19.1. *Summary impacts associated with extreme rainfall in East Africa during 1997/98 (modified from FAO/GIEWS, 1998; Conway, 2002)*

Country	Agricultural impacts	Non-agricultural impacts
Kenya	1997/98 maize crop affected during harvest. Worst effects on second-season crops (mid-October to February) – yields reduced by up to 33%.	Serious floods caused loss of life, damage to housing and infrastructure. Outbreak of Rift Valley Fever after December 1997. Beneficial effect on pastures.
Tanzania	Localised crop losses and damage to Vuli crop in lowland areas. Some gains in yield in highland areas because of higher rainfall.	Disruption of road and rail systems. Beneficial effect on pastures.
Uganda	Despite localised crop losses from flooding, the heavy rains were beneficial for crop development.	Floods and mudslides, mainly in eastern areas, caused loss of life, damage to housing and infrastructure.

rise in population inhabiting lakeshore areas and included loss of crops and livestock and destruction of homes (Conway et al., 2005; Goulden, 2006).

Vulnerability and Resilience to Flooding in Lake Shore Villages

We examined the impacts of the 1997/98 flooding event for two Ugandan villages, one on the southern shore of Lake Kyoga in Buyende district and the other on the southern shore of Lake Wamala, a small lake in the catchment of Lake Victoria, in Mityana district (see Figure 19.1 for locations and Table 19.2 for key features of the villages). Field work carried out in these villages between 2003 and 2005

Table 19.2. *Key features of the study villages*

	Lake Kyoga village	Lake Wamala village
Population	201 households	377 households
Major ethnic groups (dominant group in bold)	**Bakenyi**, Basoga, Banyoro, Itesot	**Baganda**, Banyaruanda, Basoga, Bakiga
Main livelihoods	Cultivation, livestock, fishing	Cultivation, livestock, fishing
Main crops	Sweet potatoes, millet, cassava, maize, groundnuts	Beans, cassava, sweet potatoes, maize, bananas
Closest urban centre	Kamuli town, 50 km	Mityana town, 14 km
Annual rainfall (mm)	1,475 mm (mean annual rainfall 1970 to 2002 for Kiige rainfall station)	1,191 mm (mean annual rainfall 1940 to 1985 for Mityana rainfall station)

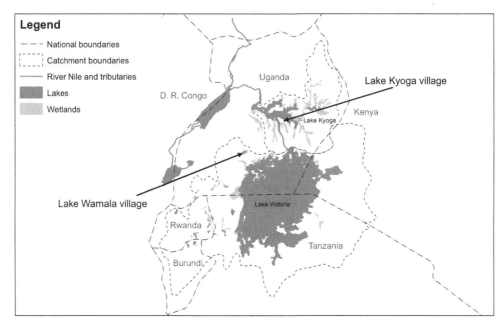

Figure 19.1 Map showing the location of the study villages and the location of lakes Kyoga and Wamala, Uganda.

examined the population's adaptive responses and factors that contributed to resilience or vulnerability to climate stresses. This involved a household livelihood survey, semi-structured interviews and group discussions (Goulden et al., 2013), supplemented by information from interviews with Ugandan policymakers and other stakeholders between 2003 and 2008. The household sample was stratified by wealth, established by wealth-ranking exercises. Household resilience was measured using an indicator of subjective well-being (Camfield and McGregor, 2005) supplemented by qualitative interview data.

Factors that affected the resilience of households to the flooding included the amount of natural, physical, financial, human and social capital assets that they were able to access, as well as their exposure to the flooding, partly governed by the location of homes and crop land. The poorest households and those headed by women and elderly people were least able to cope with the flooding impacts that included destruction of their homes and crops, death of livestock and a temporary decrease in fishing income. Their reduced resilience was because of a lower diversity of livelihood activities based on reduced labour availability, less access to higher-income-earning activities and, in some cases, reduced social capital (Goulden et al., 2013). Better-off households were able to improve their resilience to climate impacts by diversifying their livelihoods into higher-income activities less affected by climate shocks, such as trading activities that accessed more distant markets, making use of social networks and local knowledge (Goulden et al., 2009; 2013).

In the Lake Kyoga village, a large number of households were particularly vulnerable because of their limited access to assets (natural, physical, financial and social capital, particularly land) and increased exposure because of their location close to the lake shore since fleeing from conflict on the north-eastern side of the lake during the late 1980s and early 1990s. Exposure to several shocks such as conflict, drought and then flooding increased the vulnerability of many households. Few people had memory or experience of the previous severe flooding event in the early 1960s (Goulden, 2006).

Households used several types of livelihood diversification and social capital to respond to climate impacts and other stresses, some of which were more effective at providing resilience to the flooding than others (Goulden, 2006). For example, the *concurrent livelihood diversity* of households (doing many activities at the same time) was reduced as some activities failed because of the flooding (e.g. crops were destroyed and livestock lost), but households were more resilient where they had *spatial livelihood diversity*, because one household member was able to seek work elsewhere or they had planted crops both close to the lake shore and further away on higher land. *Temporal diversification* was used by households that swapped from one activity that was more affected by the flooding to another that was less affected; for example, many households that lost crops around Lake Wamala returned to fishing when the fishery recovered after the lake expanded. Those households that had access to *bridging* and *linking social capital* (connections with people outside their immediate family and neighbourhood and with more economically or politically influential people; Woolcock, 2000) were more resilient to the impacts of the flooding than those households that only had *bonding social capital* (connections among family, friends and neighbours; Woolcock, 2000). They were able to make use of *bridging* and *linking social capital* to access credit or knowledge to improve or start new livelihood activities or access distant markets for trading activities, for example (Goulden et al., 2013).

Responses to Flooding: Emergency Management and Post–Event Behaviour

Institutions such as local government departments, village councils, village-based credit groups and fishing committees played a role in both increasing and decreasing adaptive capacity and resilience to the impacts of the flooding. For example, village leaders were able to encourage people to rescue their belongings from the rising flood waters and assist people in seeking a new location to rebuild their homes during the flooding. Government institutions such as the District Fisheries, Agriculture and Environment departments had inadequate capacity to respond in terms of personnel and funding. There were also differences between the goals and values of government policies of natural resource management and the immediate interests of householders to secure their livelihoods. These factors limited the ability of these institutions to respond to emergencies in the short term and strengthen livelihoods and improve resilience in the longer term (Goulden et al., 2013). Emergency management responses by government institutions to the 1997/98 flood event in Uganda were limited at the community scale, the studied villages having received little or no emergency aid at the time of the flooding.[2] The responses described in the previous section were consequently of most importance, albeit highly limited for the most vulnerable members of the community, in a situation in which social protection measures such as 'safety-nets' in the form of cash transfers from government and access to services, credit and insurance (Davies et al., 2008; Heltberg et al., 2009) did not exist or were very limited.

The Uganda Government Ministry of Water, Lands, and Environment responded to the floods in several ways. Firstly, during 1998 up to 2000, they modified the releases from the dam at the outlet of Lake Victoria to mitigate the flooding downstream at Lake Kyoga. They then accepted the cooperation of Egypt's Ministry of Irrigation, which provided cutting equipment to unblock the outlet of Lake Kyoga, which succeeded in slowly reducing the level of Lake Kyoga.

Present-Day Vulnerability to Floods in the White Nile Basin

A visit in March 2010 to the village studied on Lake Kyoga that was affected by flooding in 1997/98 showed that little has changed for the livelihoods of the poor. There has been some improvement in infrastructure in terms of a better road and facilities for fish storage. Flooding has not caused problems in this area since 2007 and the 2007 floods were less severe in this location than those of 1997/98. However, drought remains a frequent problem that affects crop yields and exacerbates poverty in the region.

The National Adaptation Programmes of Action (NAPA) published by Uganda in 2007 noted a number of adaptations to climate stresses made by communities in Uganda but takes insufficient account of the need to build resilience by facilitating autonomous adaptation (Vincent et al., 2013). Adaptation projects proposed for funding under the NAPA relevant to managing flooding risks include the strengthening of early warning systems and dissemination of weather information[3] and establishing emergency and disaster management plans[4] (Government of Uganda, 2007). The Ugandan Government responded to the floods of 2007 by monitoring and mapping the flood's extent and coordinating the emergency response between government

---

[2] District fisheries officers participated in a flood damage assessment and the Red Cross distributed a small amount of emergency assistance to flood victims, which recipients found to be inadequate.

[3] This is listed under the Strengthening Meteorological Services Project, given third priority in the list of projects.

[4] This is listed under the Community Water and Sanitation Project, given fourth priority in the list of projects.

departments and non-governmental organisations (NGOs) (M. C. Goulden, unpublished data).

Basin-level efforts to improve cooperation through the Nile Basin Initiative (NBI), sponsored by the national governments of the eleven countries that share the basin[5] and international donors, have the potential to improve adaptive capacity for dealing with floods in the region. Examples of relevant NBI initiatives that are building adaptive capacity to respond to floods include the Flood Preparedness and Early Warning Project for the Eastern Nile Basin and the Decision Support System in the Water Resources Planning and Management Project. It remains to be seen how effectively these will be implemented given current political and economic developments in the Nile Basin (Cascão, 2009; Nicol and Cascão, 2011). The East African Community provided a forum for technical cooperation in response to the lowering of the level of Lake Victoria in 2005 and 2006 (M. C. Goulden, unpublished data) and may have the facility to respond to regional flooding events, for example through the Lake Victoria Basin Commission, but this has yet to be demonstrated.

19.3 Adaptation to Floods in Nile Basin in the Present Day: Lessons Learned

The 1997/98 and more recent flooding events in 2006 and 2007 highlight the continued vulnerability to flooding in the Nile Basin and considerable barriers to adaptation both at the household level and at the level of national and regional institutions. Efforts to strengthen transboundary cooperative management of water resources in the Basin through the Nile Basin Initiative and the East African Community provide opportunities for improved adaptive capacity but still face barriers to cooperation over water resources that are largely political and economic (Cascão, 2009; Nicol and Cascão, 2011). This is of particular concern because many climate model projections for East Africa indicate an increased incidence of extreme flooding events in the future (McHugh, 2005). The research presented here shows that those who are most vulnerable lack sufficient capacity to adapt their livelihoods to flooding. Attempts to improve livelihood resilience may therefore require additional protection in the case of damaging flood events. Increasingly, there are calls for social protection to be incorporated into policies for disaster risk reduction and climate change adaptation (Davies et al., 2008; Heltberg et al., 2009). This research makes a case for targeted support for those most vulnerable to climate-related risks. There is also a need for further research to understand the barriers to combining social protection, disaster risk reduction and climate change adaptation policies (Dovers, 2009).

---

[5] Southern Sudan joins the ten existing riparian countries in the Nile Basin Initiative from July 2011, following the referendum for secession that took place in January 2011 (Nicol and Cascão, 2011).

References

Adger, W. N. (2006). Vulnerability. *Global Environmental Change*, 16(3), 268–281.

Birkett, C., Murtugudde, R. and Allan, T. (1999). Indian Ocean climate event brings floods to East Africa's lakes and the Sudd marsh. *Geophysical Research Letters*, 26(8), 1031–1034.

Camfield, L. and McGregor, A. (2005). Resilience and well-being in developing countries. In *Handbook for Working with Children and Youth: Pathways to Resilience across Cultures and Contexts*, ed. M. Ungar. Thousand Oaks, CA: Sage Publications, pp. 189–210.

Cascão, A. E. (2009). Changing power relations in the Nile River basin: unilateralism vs. cooperation? *Water Alternatives*, 2(2), 245–268.

Christie, F. and Hanlon, J. (2001). *Mozambique and the Great Flood of 2000*. Oxford, UK: James Currey.

Conway, D. (2002). Extreme rainfall events and lake level changes in East Africa: recent events and historical precedents. In *East African Great Lakes: Limnology, Palaeolimnology and Biodiversity*, ed. E. O. Odada and D. O. Olago. Dordrecht: Springer, pp. 63–92.

Conway, D. (2005). From headwater tributaries to international river: observing and adapting to climate variability and change in the Nile basin. *Global Environmental Change*, 15(2), 99–114.

Conway, D., Allison, E., Felstead, R. and Goulden, M. (2005). Rainfall variability in East Africa: implications for natural resources management and livelihoods. *Philosophical Transactions of the Royal Society*, 363(1826), 49–54.

Davies, M., Guenther, B., Leavy, J., Mitchell, T. and Tanner, T. (2008). Adaptive social protection: synergies for poverty reduction. *IDS Bulletin*, 39(4), 105–112.

Dovers, S. (2009). Normalizing adaptation. *Global Environmental Change*, 19 (1), 4–6.

FAO/GIEWS – Food and Agriculture Organization Global Information and Early Warning System (1998). *Heavy Rains Attributed to El Niño Cause Extensive Crop Damage in Parts of Eastern Africa*. Rome: Food and Agriculture Organization Global Information and Early Warning System. Accessed 29 October 2012 from: <http://www.fao.org/docrep/004/w7832e/w7832e00.htm>.

Few, R. (2003). Flooding, vulnerability and coping strategies: local responses to a global threat. *Progress in Development Studies*, 3(1), 43–58.

Goulden, M. (2006). *Livelihood Diversification, Social Capital and Resilience to Climate Variability Amongst Natural Resource Dependent Societies in Uganda*. Unpublished PhD thesis. Norwalk, UK: University of East Anglia.

Goulden, M., Næss, L. O., Vincent, K. and Adger, N. (2009). Accessing diversification, networks and traditional resource management as adaptations to climate extremes. In *Adapting to Climate Change: Thresholds, Values, Governance*, ed. W. N. Adger, I. Lorenzoni and K. O'Brien. Cambridge, UK: Cambridge University Press, pp. 448–464.

Goulden, M. C., Adger, W. N., Allison, E. H. and Conway, D. (2013). Limits to resilience from livelihood diversification and social capital in lake social-ecological systems in Uganda. *Annals of the Association of American Geographers*, 103(4), 906–924. doi:10.1080/00045608.2013.765771.

Government of Uganda (2007). *Climate Change: Uganda National Adaptation Programmes of Action*. Kampala: Government of Uganda. Accessed 29 October 2012 from: <http://unfccc.int/resource/docs/napa/uga01.pdf>.

Gumbricht, T. (2005). Lake level and area variations 1960 to 2002 in Lake Kyoga, Uganda. In *Proceedings of the 11th World Lakes Conference*, eds. E. O. Odada, D. O. Olago, W. Ochola et al., vol. 2. Nairobi, Kenya: International Lake Environment Committee, pp. 392–395.

Heltberg, R., Siegel, P. B. and Jorgensen, S. L. (2009). Addressing human vulnerability to climate change: toward a 'no-regrets' approach. *Global Environmental Change*, 19(1), 89–99.

Jury, M. (2010). Meteorological scenario of Ethiopian floods in 2006–2007. *Theoretical and Applied Climatology*, 101, 29–40.

López-Marrero, T. and Tschakert, P. (2011). From theory to practice: building more resilient communities in flood-prone areas. *Environment and Urbanization*, 23(1), 229–249.

McHugh, M. J. (2005). Multi-model trends in East African rainfall associated with increased CO_2. *Geophysical Research Letters*, 32(1), L01707.

Moges, S., Alemu, Y., McFeeters, S. and Legesse, W. (2010). Flooding in Ethiopia: recent history and the 2006 flood. In *Water Resources Management in Ethiopia: Implications for the Nile Basin*, ed. H. Kloos and W. Legesse. New York: Cambria Press, pp. 285–305.

Nicol, A. and Cascão, A. E. (2011). Against the flow – new power dynamics and upstream mobilisation in the Nile Basin. *Review of African Political Economy*, 38(128), 317–325.

Penning-Rowsell, E., Johnson, C. and Tunstall, S. (2006). 'Signals' from pre-crisis discourse: lessons from UK flooding for global environmental policy change? *Global Environmental Change*, 16(4), 323–339.

Schipper, L. and Pelling, M. (2006). Disaster risk, climate change and international development: scope for, and challenges to, integration. *Disasters*, 30(1), 19–38.

Smit, B. and Wandel, J. (2006). Adaptation, adaptive capacity and vulnerability. *Global Environmental Change*, 16(3), 282–292.

Sutcliffe, J. V., Dugdale, G. and Milford, J. R. (1989). The Sudan floods of 1988. *Hydrological Sciences Journal*, 31, 355–364.

Tarhule, A. (2005). Damaging rainfall and flooding: the other Sahel hazards. *Climatic Change*, 72(3), 355–377.

Thomalla, F., Downing, T., Spanger-Siegfried, E., Han, G. Y. and Rockstrom, J. (2006). Reducing hazard vulnerability: towards a common approach between disaster risk reduction and climate adaptation. *Disasters*, 30(1), 39–48.

Tschakert, P., Sagoe, R., Ofori-Darko, G. and Codjoe, S. N. (2010). Floods in the Sahel: an analysis of anomalies, memory, and anticipatory learning. *Climatic Change*, 103, 471–502.

Vincent, K., Naess, L. O. and Goulden, M. (2013). National level policies versus local level realities – can the two be reconciled to promote sustainable adaptation? In *A Changing Environment for Human Security: Transformative Approaches to Research, Policy, and Action*, ed. L. Sygna, K. L. O'Brien, and J. Wolf. London, UK: Earthscan, pp. 126–134.

Walker, B., Gunderson, L., Kinzig, A. et al. (2006). A handful of heuristics and some propositions for understanding resilience in social-ecological systems. *Ecology and Society*, 11(1), 13.

WaterAid (2008). *WaterAid Uganda: Annual Report for 2007/8*. Kampala, Uganda: WaterAid.

Webster, P. J., Moore, A. M., Loschnigg, J. P. and Leben, R. R. (1999). Coupled ocean-atmosphere dynamics in the Indian Ocean during 1997–98. *Nature*, 401(6751), 356–360.

Woolcock, M. (2000). Friends in high places? An overview of social capital. *Development Research Insights*, 34. Sussex, UK: Institute of Development Studies.

20

Floods in the Yangtze River Basin, China

ZBIGNIEW W. KUNDZEWICZ, JIANG TONG, AND SU BUDA

20.1 Context

The distribution of precipitation over China is very uneven in space and time. During the flood season (of the summer monsoon, usually extending from May to October), precipitation may make up three-quarters or more of the annual total. Floods frequently devastate riparian parts of the more humid east and south of China, but they also severely hit the drier north and north-east of China.

A significant part of the global flood losses in the last decades have been recorded in China. During many individual years, the number of flood fatalities in China exceeds 1,000 (yet it was much higher in the past, e.g. 8,500 per year, on average, in the 1950s) and the level of material damage exceeds US$1 billion. According to Munich Re data (Berz and Kron, 2004), the 1998 (on Yangtze and Songhua rivers) and the 1996 (on Yangtze, Huanghe and Huaihe rivers) floods in China caused material damage of US$30.7 billion and US$24 billion, respectively (in nominal, 1998 and 1996 dollars). It is estimated that the dramatic 1998 flood in China, culminating in the south and east but also devastating the north-east, killed 5,500 people, left 21 million homeless and inundated more than 23 million hectares. There were 3,650 fatalities in the Yangtze river basin alone.

The globally highest nominal river flood damage was recorded in China in 2010 when the toll soared to US$51 billion. Using data from Chinese statistical yearbooks, one can find that flood-affected areas in China have been subject to a growing trend in the last sixty years (Figure 20.1). Since 2000, average annual economic damage caused by floods in China has been approximately 0.64 per cent of GDP, or five times less than 3.39 per cent in the 1990s (Cheng and Zhang, 2011).

Floods have been particularly commonplace in the basin of the Yangtze River (*Changjiang* in Chinese), the largest one in China, with more than 6,397 km in length and an area of 1.8 million km^2. The river flow and the water level of the Yangtze River and its tributaries vary in a broad range. Destructive abundance of water continues to

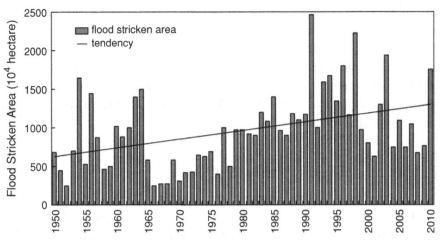

Figure 20.1 Flood-affected area in China for 1950–2010.

play havoc on the riparian population of hundreds of millions of people. There have been numerous dramatic floods in the Yangtze system since the dawn of civilisation until recently.

Floods on the Yangtze River are triggered by monsoon rainfalls, which differ in intensity, duration, areal extent and timing. A regular feature in the Yangtze Basin is the occurrence of *mei-yu* (in Chinese: plum rain) season, during June and July (Jiang et al., 2008). The plum rain season, coincident with the period when the fruit matures, usually begins in mid-June and ends approximately twenty days later. Typically it is caused by a quasi-stationary rain belt covering the middle/lower Yangtze Basin, bringing in abundant precipitation. The onset and length of *mei-yu*, which largely drives the flood occurrence, is likely to change because of changes in monsoon circulation. The *mei-yu* period may come earlier and last longer.

Flood risk changes depend on several factors, climate being one, while land use change, urbanisation and human encroaching into flood plains have been also important. Accelerated urbanisation can be observed in many parts of China. A particular example is the Yangtze River Delta Region covering 1 per cent of the area of China, housing 6 per cent of the total population and producing 20 per cent of the nation's GDP.

20.2 Floods on the Yangtze in the Last 150 Years

In the last 150 years, seven extraordinarily heavy floods occurred on the Yangtze and its tributaries in 1860, 1870, 1931, 1935, 1954, 1998 and 2010, leading to high death tolls, material damage and social losses. In 1870, the highest Yangtze discharge in the observation period was recorded in Yichang (105,000 m^3 s^{-1}) and also a thirty-day river discharge reached a record level of 165 billion m^3. This was likely to be the largest flood on the Yangtze during the last millennium. In the twentieth century, dramatic destruction was caused by floods in 1931 and 1954.

A particular intensification of material flood damage was observed since 1990. There were six large floods in the 1990s, namely in 1991, 1993, 1995, 1996, 1998, and 1999. The exceptional 'density' of floods in the 1990s seems to indicate an increase in the frequency of large floods. This tendency continued in the first decade of the twenty-first century.

The 1998 Floods

The destructive floods on the Yangtze in the summer of 1998 and its tributaries (and on the other rivers in China) illustrate the typical flood-generating mechanism. Intense and long-lasting precipitation spells started already in May. Then *mei-yu* rainfalls in late June and early July led to the formation of high flood peaks and the subsequent heavy rainfall aggravated the situation, causing a prolonged flood persisting for several months. In the summer of 1998, a stable rain belt stayed for several weeks over the Yangtze river basin. On some days, a twenty-four-hour precipitation exceeding 100 mm occurred over large areas, in excess of 50,000 km². From June to August, there were eighty-four days of storm rain with more than ten stations recording daily precipitation in excess of 50 mm in each of them and fifty-two days of heavy rainstorm (with more than ten stations with daily rainfall over 100 mm). Daily precipitation in excess of 300 mm/d was noted in six days.

The 2010 Floods

The most recent destructive floods, landslides, gully erosion and debris flows occurred in 2010, spanning the interval from May until late October. The monsoon rainfalls lasted long and, locally, were very intense. In Guangdong Province, a rainfall of 603.5 mm fell within six hours. The disaster had a large-scale areal coverage. Floods and landslides affected twenty-eight provinces, autonomous regions and municipalities, including dry-climate areas, such as the Xinjiang Province in north-west China. The damage over China, extending beyond the Yangtze river basin, was very severe, with more than US$51 billion in material damage, 3,200 fatalities and 1,000 missing. The floods and landslides affected more than 230 million people – on average, every sixth citizen of China. More than 16.5 million hectares of farmland were affected and more than 2 million hectares completely destroyed. The Gansu mudslide of 8 August killed more than 1,500 (Cheng and Zhang, 2011).

In 2010, the Three Gorges Dam on the Yangtze was subject to a difficult test of 70,000 $m^3 s^{-1}$ discharge (more than the discharge generated in the upper Yangtze in 1998), out of which 40,000 $m^3 s^{-1}$ was released and 30,000 $m^3 s^{-1}$ was held back. The reservoir level went up to 158.86 m (with alarm level being 145 m), and a 4 m increase in the water level occurred overnight. Even at the main gauging stations in the middle and the lower reaches, water levels were above the alarm level.

20.3 Climate Change Track – Observations and Projections

One can identify a possible climate track in floods in China by analysing observations and model-based projections of precipitation, intense precipitation and river discharge, with consideration of the multi-factor context.

China Climate Change Bulletin (China Meteorological Administration, 2010) illustrates decreasing trends in the annual number of rainfall days and increasing annual accumulated number of heavy rain-station days in 1961 through 2010. It also shows that the variability of frequency of regional heavy rainfall events in China is on the rise. Qian and colleagues (2007) illustrate strong variability of the climate regime in time scale of decades.

Gemmer and colleagues (2011a) review changes in heavy precipitation in China, reporting on positive trends in annual maximum precipitation in the Yangtze river basin, and increase in above normal mean intensity in east China. Chen and colleagues (2011) analysed variations of extreme precipitation for 1956 through 2008, corroborating earlier findings on complex changes. Increases of maximum daily precipitation and in the number of annual precipitation days were found in much of south of China, but some changes (e.g. on the Yangtze) were not statistically significant at 0.05 level. Both the annual and extreme precipitations (above the 95th percentile) have shown an upward trend in recent decades in the middle and lower Yangtze reaches.

The summer precipitation in the Yangtze river basin has evidenced an upward trend in recent decades, especially in June and July (Jiang et al., 2008). The upward precipitation trend in July covers the whole basin, and in June it prevails in the middle and the lower regions. The increase of monthly precipitation in the *mei-yu* season, June and July, is of much importance for flood hazard. Influenced by the increasing summer precipitation and rainstorm trends, the summer runoff and flood discharge of the lower Yangtze region show an upward trend in the recent decades (Jiang et al., 2008).

Since the low-lying and broad valley of the lower part of the Yangtze river basin is densely populated, the region is very vulnerable to flooding and inundations cause very high material damage. If the presently observed trend continues into the future, higher flood risk in the lower part of the basin can be expected. Climate projections show warming for all seasons, and most models project wetter conditions over much of China. This also pertains to the principal flood season, June to August. Hirabayashi and colleagues (2008) compared the flood risk projections in the horizon 2071 through 2100 and the control period. What used to be a 100-year flood in China in the control period is likely to occur much more frequently in the future, under changed climate, with return a period of 50 years and below.

There is no doubt that flood risk has grown in many places in China and is likely to grow further in the future because of a combination of anthropogenic and climatic factors. Intense precipitation grows in the warming climate. However, reliable and detailed quantification of aggregate flood statistics is very difficult to obtain for the past to present and is virtually impossible to obtain for the future.

20.4 Adaptation – Lessons Learned

Flood protection, or adaptation to natural variability of discharge, has been developed in China for millennia. The level of expenditure on flood protection in China has grown considerably in recent decades. However, even if there exist powerful embankments along the Yangtze River and its many tributaries, they do not provide satisfactory protection of the riparians during large floods (cf. Kundzewicz and Xia, 2004). Increasingly, large flood damage has been recently occurring on medium- and small-size rivers in the Yangtze Basin.

Despite massive efforts, it is getting abundantly clear that complete flood control is not possible. Disastrous floods frequently plague the country. Hence, flood risk management is needed in China. After a suite of destructive floods in China, increasing attention is being paid to upgrading flood protection measures. The country has embarked upon an ambitious and vigorous task to improve flood preparedness by both structural ('hard') and non-structural ('soft') measures. The former refer to such defences as dikes, dams and flood control reservoirs, diversions and the like. The latter include implementing watershed management (source control), zoning, insurance, flood forecasting-warning systems and awareness raising.

Structural measures, both dikes and dams of different sizes, have a very long tradition in China and play a vital role in flood prevention also today. By 2002, more than 86,000 reservoirs had been built in China, with a total storage capacity of more than 0.5×10^{12} m^3, accounting for more than one-fifth of the total estimated annual runoff from the land areas (Guo et al., 2004). Most reservoirs serve multiple purposes: Flood control, hydropower, irrigation, water supply, navigation and so on. The total number of dams has increased very strongly in the last fifty years. In 1960, there were only five large dams (taller than 100 m), but their number increased ten-fold to fifty in 2000 (Xu et al., 2010).

The multi-objective, massive Three Gorges Dam on the Yangtze, the world's greatest engineering work, has flood protection as the principal objective. In the last fifty years, more than 200,000 km of dikes have been strengthened for alleviating the impacts of floods in the country (Zhang et al., 2002).

In addition to the traditional structural measures, it is increasingly recognised in China that non-structural measures deserve considerable interest. Essential advances in flood forecasting systems have been made possible by the integration of mathematical modelling, advanced information technology, telecommunication and remote sensing (Zhang et al., 2002). Integrated approaches enable increasing flood warning lead-time and, in turn, can reduce flood damage and save lives.

After the disastrous 1998 flood, the Chinese Government has intensified efforts to strengthen flood defences – to reinforce dikes and dredge river channels in order to improve flood conveyance. It was attempted to stop reclamation, thus restoring polders and increasing storage capacity, to relocate residents of flood plains and to convert cultivation to reforestation.

The Chinese Government is implementing plans to build national weather radar networks and to use radar data for real-time flood forecasting in order to improve accuracy of flood forecasting. Coupling of radar data with hydrological models is carried out, overcoming differences in spatial resolutions of the remotely sensed data and the hydrological models.

Gemmer and colleagues (2011b) reviewed climate change adaptation in China, the National Climate Change Program and China's White Paper 'China's Policies and Actions for Addressing Climate Change'. By 2009, all thirty-four provinces produced a climate change plan. The Flood Prevention Law (2007) lays out principles and responsibilities for flood prevention planning. There is a national standard (GB50201–94) drafted by the Ministry of Water Resources and issued by the Ministry of Construction in 1994 dealing with flood return periods for different types of locations (Gemmer et al., 2011b). In 2010, flood hazard mapping guidelines were published as professional standards by the Ministry of Water Resources. This was based on experience collected by way of mapping exercises carried out in pilot studies since 2005.

Since the information of relevance to flood protection is distributed among several agencies, effective cooperation and communication among federal, state and local stakeholders is essential. This is inherently difficult, but progress has been achieved in flood forecasting integration, data sharing and collaborative problem solving. The China Meteorological Administration (CMA) collects observations of precipitation and other meteorological variables and prepares precipitation forecasts. The Ministry of Water Resources (MWR) collects hydrological observations (e.g. of river levels and discharges) and is responsible for flood forecasting and dissemination of the forecast. River Basin Commissions (altogether seven, including the Yangtze River Basin Commission) are agencies of the MWR.

20.5 Gaps in Knowledge and Concluding Remarks

Studying changes that influence flood hazard and flood risk in China is a challenging area. Possibly, there are changes in intense precipitation, changes in cyclone track, changes in land use and changes in exposure and vulnerability. Early detection and attribution of changes would be of vast practical importance.

The flood hazard depends on a combination of anthropogenic and natural factors, such as climate, land use and population density. The total population of China increased from 560 million in the 1950s to 1,300 million in 2000. The Yangtze river basin is densely populated (more than 400 million inhabitants) and largely urbanised. Owing to the growing population pressure, activities like deforestation, agricultural land expansion, urbanisation, construction of roads and reclamation of wetlands and lakes have been progressing. This has reduced the available storage capacity in the basin, increased the value of the runoff coefficient and aggravated the flood hazard.

Floods on the Yangtze may considerably differ in areal coverage, ranging from local and regional up to basin-wide events. Floods are particularly dramatic if superposition of flood waves occurs – for example, when high waves from the Dongting and Poyang lake basins meet with a flood wave on the mainstream Yangtze and when the flood peaks from the upper and the middle/lower reaches coincide, causing a gigantic flood wave crest downstream.

References

Berz, G. and Kron, W. (2004). Überschwemmunskatastrophen und Klimaänderung: trends und Handlungsoptions aus (Rück-) Versicherungssicht (Flood disasters and climate change: trends and options – a (re-) insurance view, in German). In *Warnsignal Klima: Genug Wasser für alle?*, ed. J. L. Lozan, H. Graßl, P. Hupfer, L. Menzel and C.-D. Schönwiese. Hamburg, Germany: Wissenschaftliche Auswertungen, pp. 264–269.

Chen, Y., Chen, X. and Ren, G. (2011). Variation of extreme precipitation over large river basins in China. *Advances in Climate Change Research*, 2, 108–114.

Cheng, X. and Zhang, D. (2011). Recent trend of flood disasters and countermeasures in China. In *Large-scale Floods Report*, ed. A. Chavoshian and K. Takeuchi. Ibaraki-ken, Japan: ICHARM, pp. 189–195.

China Meteorological Administration (2010). *China Climate Change Bulletin*. Beijing, China: China Meteorological Administration.

Gemmer, M., Fisher, T., Jiang, T., Buda, S. and Lü, L. L. (2011a). Trends in precipitation extremes in the Zhujiang River Basin, South China. *Journal of Climate*, 24, 750–761.

Gemmer, M., Wilkes, A. and Vaucel, L. M. (2011b). Governing climate change adaptation in the EU and China: an analysis of formal institutions. *Advances in Climate Change Research*, 2, 1–11.

Guo, S., Honggang, Z., Hua, C. et al. (2004). A reservoir flood forecasting and control system for China. *Hydrological Sciences Journal*, 49, 959–972.

Hirabayashi, Y., Kanae, S., Emori, S., Oki, T. and Kimoto, M. (2008). Global projections of changing risks of floods and droughts in a changing climate. *Hydrological Sciences Journal*, 53, 754–773.

Jiang, T., Kundzewicz, Z. W. and Buda, S. (2008). Changes in monthly precipitation and flood hazard in the Yangtze River Basin, China. *International Journal of Climatology*, 28, 1471–1481.

Kundzewicz, Z. W. and Xia, J. (2004). Towards an improved flood preparedness system in China. *Hydrological Sciences Journal*, 49, 941–944.

Qian, W., Lin, X., Zhu, Y., Xu, Y. and Fu, J. (2007). Climatic regime shift and decadal anomalous events in China. *Climatic Change*, 84, 167–189.

Xu, K., Milliman, J. D. and Xu, H. (2010). Temporal trend of precipitation and runoff in major Chinese Rivers since 1951. *Global and Planetary Change*, 73, 219–232.

Zhang, J. Q., Zhou, C. H., Xu, K. Q. and Watanabe, M. (2002). Flood disaster monitoring and evaluation in China. *Environmental Hazards*, 4, 33–43.

Part V

Synthesis Chapters

21

Disasters and Development

JESSICA AYERS,* SALEEMUL HUQ, AND SARAH BOULTER

Climate-related disasters are defined by development: Poor people are disproportionately affected (Pelling, 2003; Wisner et al., 2004; Ribot, 2010). In the first place, natural hazards are unevenly distributed. In 2007, Asia was the region hardest hit and most affected by natural disasters, accounting for 37 per cent of reported disasters and 90 per cent of all the reported victims (Dodman et al., 2009). Human-induced climate change impacts such as sea-level rise and increased climate extremes are likely to have the heaviest impact on small island developing states, the poorest countries in the world and African nations (IPCC, 2007).

Development context determines vulnerability to natural hazards. For example, individuals and households that have reliable access to food and adequate food reserves, clean water, health care and education will inevitably be better prepared to deal with a variety of shocks and stresses – including those arising because of climate change (Dodman et al., 2009). The role of development in determining the risk posed by natural hazards is now well established in the disaster-risk reduction literature (Anderson and Woodrow, 1989; Wisner et al., 2004; Kelly and Adger, 2009).

But what does this relationship mean for managing adaptation to climate-related disasters? Does the role of development in defining climate-related risks have implications for how those risks are (or should be) responded to? Should climate disasters be managed differently in developed and developing countries? This chapter addresses these questions with reference to the case studies within this book and elsewhere.

21.1 The Relationship between Development and Climate-Related Disasters

The role of development in defining disasters is underpinned by the concept of 'vulnerability'. 'Vulnerability' is broadly understood as 'being prone to or susceptible to damage or injury' (Wisner et al., 2004: 9).

* At the time of writing, Jessica Ayers was affiliated with the International Institute for Environment and Development.

199

However, the extent to which this matters to different groups is determined by factors related to their socio-economic development. Large landowners may have more efficient irrigation and drainage systems than small farmers; wealthier households may have the means (capacity, education, resources) to diversify their income options during the wet season; poorer households may be more exposed to injury and have fewer means to access healthcare systems during times of crisis. This is summed up by Ribot (2010) in the following statement:

> The poor and wealthy, women and men, young and old, and people of different social identities or political stripes, experience different risks while facing the same climatic event... the inability to manage stress does not fall from the sky. (Ribot, 2010: 49)

From this perspective, the role of development is not in defining exposure to a specific hazard but in enabling or undermining adaptive capacity in order to build resilience to a range of stresses including climatic hazards. Development defines the 'drivers' or underlying causes of vulnerability that determine not only whether people are exposed in the first place but also whether they are able to cope with and adapt to that exposure.

21.2 The Role of Development in Shaping Climate-Related Disaster Risk Reduction and Response

Different frameworks for understanding vulnerability shape the role of development in disaster response and in climate change adaptation. Early attempts to develop guidelines for mitigating disasters resulted in conventional disaster management approaches that specifically targeted the impacts of hazards, with policy recommendations for managing disasters focusing on narrowly technological engineering approaches to controlling the physical environment (Pelling, 2001; Wisner et al., 2004). This approach demanded solutions designed by a small team of highly technical experts that could analyse the hazard and develop targeted technical responses.

However, during the 1980s, many observers from disaster risk reduction and development studies began to draw attention to the link between the risks people face and the reasons behind their vulnerability to these risks in the first place (Sen, 1981; Hewitt, 1983; Anderson and Woodrow, 1989; Blaikie et al., 1994). Since then, three decades of work in disaster risk reduction have highlighted the ways in which technological approaches to risk management have focused consultations on expert opinion to the exclusion of stakeholders and communities set to benefit from the outcomes (Pelling, 2001).

This recognition led to a shift in disaster management in developing countries that began to recognise and address factors related to under-development as a first step in building resilience to disasters. This, in turn, gave rise to development-focused solutions and locally appropriate, livelihoods-based support in building resilience from the bottom-up.

This latter approach demands a more participatory approach to defining local risks and developing locally appropriate responses. Taking a 'social vulnerability'-based perspective on climate change risk shifts the emphasis of risk assessment away from climate change impacts and towards the local circumstances of vulnerability. This recognition has led proponents of a social vulnerability approach to argue that risk assessments that inform risk-reduction policies need to be more locally responsive and therefore inclusive (Huq et al., 2004; Few et al., 2007; Ayers, 2011). If the factors that determine vulnerability are context specific, designing risk-reduction interventions to address these factors requires a knowledge base that is tailored to local settings. Participatory and locally driven responses are therefore not only ethical but also practical when development priorities are taken as the starting point for risk reduction (Few et al., 2007). This perspective gave rise to community-based disaster risk-reduction approaches in under-developed contexts that prioritise locally appropriate development interventions designed with the participation of vulnerable communities. This approach is now well established as one of the pillars of human security within the Disaster Risk Reduction (DRR) field (Pelling, 2003).

This approach is in contrast to the early paradigm of climate change that focused on the impact of biophysical change in the atmosphere rather than the factors that make people vulnerable to these changes. 'Adaptation' to climate change emerged from this context to deal with the impacts of non-mitigated greenhouse gas emissions, resulting in an 'impacts-based' approach to managing climate change risk (Burton et al., 2002). The primary adaptation focus was 'technology-based' interventions such as dams, early warning systems, seeds and irrigation schemes based on specific knowledge of future climate conditions (Klein, 2008).

This impacts-based approach shifts the balance of climate-related disaster management back towards a hazard-risk framework that implies a particular type of scientific or technological expertise is needed to assess climate risks for policymaking. The role of 'development' in this sense is building technological and scientific capacity and is at odds with the bottom-up and locally inclusive approaches adopted in DRR. Yet the impacts-based approach has dominated climate change adaptation management under the United Nations Framework Convention on Climate Change (UNFCCC) (Huq and Toulmin, 2006; Schipper, 2006; Ayers and Dodman, 2010).

This perception of adaptation is starting to change, driven largely by the development and disaster risk reduction communities. In 2002, a report released by ten leading development-funding agencies stated that climate change was a threat to development efforts and poverty reduction, including the achievement of the Millennium Development Goals, and that pro-poor development was key to successful adaptation (Sperling, 2003). The report reflects many of the themes emerging in the DRR and vulnerability literature on vulnerability at the time (e.g. Smit et al., 2000; Huq et al., 2002; Kates, 2009), including recommendations to support sustainable livelihoods, improve governance and make institutions more accountable and participatory (Sperling, 2003;

Klein, 2008). Since 2002, research and non-governmental organisation (NGO) communities have increasingly incorporated climate change within their development work, believing they have the skills, experience, local knowledge and networks to undertake locally appropriate vulnerability-reduction activities that increase resilience to a range of factors that include climate change (Ayers and Dodman, 2010).

This shift is particularly evident through the recently emerging discourse of community-based adaptation (CBA). CBA operates outside UNFCCC–led processes, starting at the community level to identify, assist and implement community-based development activities that strengthen the capacity of local people to adapt. Proponents of CBA suggest that done well, CBA presents an opportunity for shifting the balance of risk assessments back towards participatory, locally responsive adaptation planning (Jones and Rahman, 2007; Ayers and Forsyth, 2009). Other observers point to the relative infancy of CBA and the need for more critical engagement with the learning from grass-roots development and disaster-risk reduction about how to achieve meaningful participatory action (Dodman and Mitlin, 2011).

Although this shift towards more development-orientated climatic disaster response is relatively recent, this chapter proposes that in low-income contexts, the role of development in achieving resilience has been explicitly recognised, and several of the case studies in this book highlight this. This creates entry points for incorporating vulnerable people in making more holistic choices about managing disasters, improving their own adaptive capacity and moving away from impacts-focused and expert-driven decision making. In the remainder of this chapter, we use the book's case studies and two additional case studies to illustrate how attention to the role of development can provide entry points into more effective, inclusive planning for climate-related risk reduction.

21.3 Managing Disasters in Developed and Developing Country Contexts: A Review of Case Studies

The 'Impacts-Based' Approach to Risk Management in Developed Countries

This chapter has proposed that adopting an 'impacts-based' approach to managing risks can result in the sidelining of development-related or vulnerability issues in managing disaster responses and an over-emphasis on technocratic expertise in defining 'solutions'. This is well illustrated by the historical response to flooding in the Mississippi river basin, North America (Chapter 4). In this case, enormous amounts of financial and engineering resources have been poured into building then raising and extending a system of levees and flood protection systems on the floodplains of the Mississippi River starting as far back as the early 1700s. Yet with almost ridiculous regularity, flooding events overtop or breach these structures, causing widespread inundation, death and destruction. The key flaw here is that as quickly as the flood defences are increased, the river is further constricted, the floodplain further urbanised – counteracting any 'advancement' in flood protection. The approach highlights an impacts-based approach and, in this case, one that has had limited success in real terms for the affected community.

Responses to windstorms/cyclones/hurricanes (Chapters 3, 9 and 11), in developed countries at least, are commonly technical engineering responses with a focus on building tolerance and event return periods. In the case of cyclone Tracy, a very successful adaptation of building regulations was evidenced. While the long-term results were highly successful, there were a number of important social issues that were not particularly well addressed, with some evidence that demographic changes in Indigenous populations may have resulted from the disaster (Haynes et al., 2011). In the case of hurricane Katrina (Chapter 3), it is clear that major social issues need to be addressed to reduce vulnerability in the future.

The Economist (2012) recently argued that the Dutch are reconsidering their philosophy of flood control through building ever-higher dykes to improving resilience. In the Room for River project, land use of vulnerable flood plains is being returned (in a planning sense) to floodplains through moving farms or raising buildings (*The Economist*, 2012).

We add one further case study example to our discussion. Following an unusually powerful thunderstorm in which 196 mm of rain fell in four hours at the head of a catchment, the small Cornish village of Boscastle suffered major damage. Government and Environment Agency officials framed the event as an indicator of climate change that could have severe implications for the future of the tourism industry. Based on this understanding, the Government commissioned external hydrological and climate systems experts to assess the impact of the flood and make a judgment on the appropriate solution. The resulting policy response was an expensive, highly technical engineering intervention. In her review of the process, Jennings (2009) suggests many locals viewed the policy with scepticism and even derision. Jennings argues that from the perspective of Cornish residents, the 2004 flood was the result of inept government land management practices as much as it was extreme weather events. While local residents acknowledged the role of extreme weather events on their local livelihoods and economy, they felt that assumptions about the role of climate change overshadowed the more important historical and institutional factors that had led to their dependency on a climate-sensitive tourism industry (Jennings, 2009: 247). Jennings suggests that despite apparent widespread efforts to ensure participation in decision making around policy responses to the event, knowledge perceived as 'local' was subordinated in favour of externally generated expertise.

This final case study shows how an over-emphasis on the impacts of the hazard meant that the role of development in defining vulnerabilities was sidelined and participation in determining adaptive solutions was restricted. On the contrary, Jennings shows how greater attention to lay and non-expert experiences can reveal locally embedded understandings of perceptions and experiences of risk that can allow more locally relevant risk-reduction solutions (Jennings, 2009).

Disaster Risk Management in Developing Countries

A clear and disproportionate human toll is recorded for almost all the case studies of events occurring in developing countries in this book. Although it is almost impossible to compare individual events quantitatively, it is clear that in developing countries, natural disasters often result in deaths on the order of several thousand people rather than the hundreds seen in developed countries. Disaster response often relies on assistance and aid from foreign countries and NGOs. The geographical vulnerability of developing countries is discussed in the opening of this chapter, but also of great significance are poverty, education and physical vulnerability (e.g. shelter) to extreme events.

While vulnerability is clearly tied to poverty – regardless of the economic development status of a country – the response to disasters in developing countries is very different to that in developed countries.

Of the six case studies featured in this book, we identify several commonalities. First is the community-level response that builds resilience. In Mexico, for example, community-based industry (coffee production) can be a source of resilience – with a capacity for self-organisation and learning that is helping these farmers adapt their business and communities to disasters and change (Chapter 18). Similarly, in Sahel and along the Nile, Batterbury and Mortimore (Chapter 15) and Goulden and Conway (Chapter 19) all identify the building of resilience at the community level through self-organisation and diversity.

The second commonality of the case studies in developing countries is that in many of these cases, disaster creates an entry point for development-based and participatory approaches to disaster response. These opportunities are often missed in high-income contexts in which a greater expectation is placed on the government to repair damage and return to the status quo. For example, following cyclone Sidr, the Bangladesh Government sought to adopt a 'build back better' approach to disaster recovery (Chapter 17).

In a further example, discussed in Rahman and colleagues (2009), a community-based adaptation project in the increasingly flood-prone Char Islands of the Gaibanda District, Northern Bangladesh, focused on identifying what made these flooding events 'disastrous' rather than on the biophysical impacts. While the project is a good example of community participation, it does highlight some of the challenges. The project revealed the reduced reliability of traditional knowledge systems to deal with changed weather conditions. It also highlighted the challenge of communicating complex messages about climate and change to the community. In this case, the information was not always seen as relevant to the community.

These case studies demonstrate the value of taking livelihoods and development as the starting point and emphasising participation and local capacity building in designing climate disaster management strategies.

21.4 Discussion and Conclusions

While we have highlighted the success of bottom-up approaches in the case studies mentioned in this chapter, we acknowledge that this not always the approach in developing countries, nor is it the only answer. Indeed, much adaptation planning in developing countries has been criticised for being equally top down, with an over-reliance on technical information at the expense of attention to the development-related factors that drive vulnerability (Burton, 2004; Schipper, 2007; Ayers, 2011).

Rather, we highlight cases in which development is taken as a starting point for building resilience to climate-related disasters. This provides an entry point for holistic and participatory approaches to vulnerability assessment and adaptation planning that is more likely to meet the needs of vulnerable people on the ground. We suggest that perhaps such approaches are more likely to be undertaken in low-income contexts for three reasons:

Firstly, development is perhaps a more obvious starting point in low-income contexts. Burton (2004) suggests that analysing vulnerable communities in low-income contexts would reveal an existing 'adaptation deficit', which is the existing capacity of many vulnerable countries and groups to cope with and adapt to *existing* climate risks and that any climate change adaptation program would need to reduce this deficit before those communities can adapt to future climatic changes.

The same is of course true in high-income contexts. A case in point is the devastation caused by hurricane Katrina in 2005 to New Orleans, which fell disproportionately on poor and marginalised communities. Following the disaster, there was widespread recognition of the apparent neglect of poor and/or African American citizens who tended to live in the areas most vulnerable to the flooding. Reports from the time suggested that any of the city's poor could not afford to heed hurricane warnings and flee before the hurricane struck (Chapter 3), making the disaster as much an issue of development as of the hurricane itself. And yet, part of the international outrage at the management of the disaster was precisely that Katrina exposed this kind of vulnerability in such a high-income country. Thus, development may not be an obvious starting point in developed nations, but it is nevertheless applicable.

Second, there is a long history of participatory development and disaster risk reduction in low-income countries. The importance of local participation in decision making around development interventions arose from a recognition that the managerial-style approaches of the 1970s and 1980s, dominated by professional expertise and bureaucratic control, were failing to achieve significant improvements in the livelihoods of the world's poor (Cornwall, 2000). Such observations gave rise to a 'participatory turn' in development studies and practice, emerging from the NGO community but rapidly being taken up by government and international development agencies (Williams, 2004). The trend towards more participatory approaches to development has resulted in decades of research and advocacy into locally inclusive approaches to doing development and managing disasters. Climate-related disaster management in both high- and low-income contexts would do well to draw more lessons from this experience, but perhaps the parallels are more easily made in developing countries, where the participatory development history is stronger.

Thirdly, technological and financial capacity is greater in higher-income contexts. This drives a tendency towards technical assessment and high-cost infrastructure-based solutions that are simply not an option in some low-income contexts. We have argued that prioritising high-tech, high-scale and high-cost solutions risks missing the factors that undermine adaptive capacity or risk providing a solution to the wrong problem (Handmer, 2009).

Responding to climate-related disasters may in some cases represent a practical means of achieving sustainable development – with good (i.e. sustainable) development policies and practices bolstering adaptive capacity and adaptation to climate change often meaning good development (Huq and Ayers, 2007).

Of course, critics of this approach may suggest that the role of hazards in defining risk could become too marginalised, proving problematic particularly for practical issues of governance and finance. If climate change adaptation is simply good development, what makes it adaptation? Importantly, it is argued that much existing development will become unsustainable under changing climatic conditions, so 'development as usual' is not enough in light of a changing climate context. For example, investing in roads and communication infrastructure in coastal areas would encourage settlement in those areas; however, sea-level rise may mean that such settlements will be untenable in the long term. So there is an important process of ensuring that the vulnerabilities of development as frequently evidenced in the case studies of developed countries in this book (e.g. urbanisation of floodplains) are not repeated in a development approach to adapting to climate change.

It is not proposed here that all climate-related disaster management is development; but it is suggested that it is the development context that determines how 'disastrous' a climate hazard turns out to be. Development does not cause disasters, but it certainly has a key role in defining them. As such, while development is not the same as disaster response, it is a good place to start.

Acknowledgements

We would like to thank Dr David Dodman, IIED, for discussions that informed this chapter. This chapter also draws on the presentations and discussions from Panel 5 at the 5th International Community-based Adaptation Conference, Dhaka, 2011, Building Synergies between Disaster Risk Reduction (DRR) and Community-based Climate Change Adaptation (CBA), chaired by Marcus Oxley, Global Network of Civil Society Organisations for DRR and CBA.

References

Anderson, R. M. and Woodrow, P. J. (1989). *Rising from the Ashes: Development Strategies in Times of Disaster*. Boulder, CO: Westview Press.

Ayers, J. (2011). Resolving the adaptation paradox: exploring the potential for deliberative adaptation policy making in Bangladesh. *Global Environmental Politics*, 11(1), 62–88.

Ayers, J. and Dodman, D. (2010). Climate change adaptation and development: the state of the debate. *Progress in Development Studies*, 10(2), 161–168.

Ayers, J. and Forsyth, T. (2009). Community-based adaptation to climate change: strengthening resilience through development. *Environment*, 51(4), 22–31.

Blaikie, P., Cannon, T., Davis, I. and Wisner, B. (1994). *At Risk: Natural Hazards, People's Vulnerability, and Disasters*. London, UK: Routledge.

Burton, I. (2004). *Climate Change and the Adaptation Deficit. Adaptation and Impacts Research Group Occasional Paper 1*. Gatineau, Quebec: Environment Canada.

Burton, I., Huq, S., Lim, B., Pilifosova, O. and Schipper, L. (2002). From impacts assessment to adaptation priorities: the shaping of adaptation policy. *Climate Policy*, 2, 145–159.

Cornwall, A. (2000). *Beneficiary, Consumer, Citizen: Perspectives on Participation for Poverty Reduction. Sida Studies 2*. Stockholm, Sweden: Swedish International Development Cooperation Agency.

Dodman, D. and Mitlin, D. (2011). Challenges to community-based adaptation: discovering the potential for transformation. *Journal of International Development*, 23(3), doi: 10.1002/jid.1772.

Dodman, D., Ayers, J. and Huq, S. (2009). Building resilience. In *State of the World 2009: Into a Warming World*, ed. Worldwatch Institute. Washington, DC: Worldwatch Institute, pp. 151–168.

Economist, The (2012). Counting the cost of calamities. *The Economist*, 14 January 2012. Accessed 17 January 2013 from: <http://www.economist.com/node/21542755.

Few, R., Brown, K. and Tompkins, E. (2007). Public participation and climate change adaptation: avoiding the illusion of inclusion. *Climate Policy*, 7, 46–59.

Handmer, J. (2009). Adaptive capacity: what does it mean in the context of natural hazards? In *The Earthscan Reader on Adaptation to Climate Change*, eds. L. Schipper and I. Burton. London, UK: Earthscan, pp. 213–217.

Haynes, K., Bird, D.K., Carson, D., Larkin, S. and Mason, M. (2011). *Institutional Response and Indigenous Experiences of Cyclone Tracy*. Report to National Climate Change Adaptation Research Facility, Gold Coast, Australia, June 2011.

Hewitt, K. (ed.) (1983). *Interpretation of Calamity: From the Viewpoint of Human Ecology*. Boston: Allen.

Huq, S. and Ayers, J. (2007). *Critical List: The 100 Nations Most Vulnerable to Climate Change. Sustainable Development Opinion*. London, UK: International Institute for Environment and Development.

Huq, S. and Toulmin, C. (2006). *Three Eras of Climate Change. Sustainable Development Opinion*. London, UK: International Institute for Environment and Development.

Huq, S., Reid, R., Konate, M., Rahman, A., Sokona, Y. and Crick, F. (2004). Mainstreaming adaptation to climate change in least developed countries (LDCs). *Climate Policy*, 4, 25–43.

Huq, S., Sokona, Y. and Najam, A. (2002). *Climate Change and Sustainable Development Beyond Kyoto*. London, UK: International Institute for Environment and Development.

IPCC – Intergovernmental Panel on Climate Change (2007). Summary for policymakers. In: *Climate Change 2007: Impacts, Adaptation and Vulnerability. Contribution of Working Group II to the Fourth Assessment Report of the Intergovernmental Panel on Climate Change*, eds. M. L. Parry, O. F. Canziani, J. P. Palutikof, P. J. van der Linden and C. E. Hanson. Cambridge, UK: Cambridge University Press, pp. 7–22.

Jennings, T. L. (2009). Exploring the invisibility of local knowledge in decision-making: the Boscastle Harbour flood disaster. In *Adapting to Climate Change: Thresholds, Values, Governance*, eds. W. M. Adger, I. Lorenzoni and K. L. O'Brien. Cambridge, UK: Cambridge University Press, pp. 240–253.

Jones, R. and Rahman, A. (2007). Community-based adaptation. *Tiempo*, 64, 17–19.

Kates, R. (2009). Cautionary tales: adaptation and the global poor. In *The Earthscan Reader on Adaptation to Climate Change*, eds. L. Schipper and I. Burton. London, UK: Earthscan, pp. 283–294.

Kelly, P. M. and Adger, W. N. (2009). Theory and practice in assessing vulnerability to climate change and facilitating adaptation. In *The Earthscan Reader on Adaptation to Climate Change*, eds. L. Schipper and I. Burton. London, UK: Earthscan, pp. 161–186.

Klein, R. T. J. (2008). Mainstreaming climate adaptation into development policies and programmes: a European perspective. In *Financing Climate Change Policies in Developing Countries*, ed. European Parliament. Brussels: European Parliament.

Pelling, M. (2001). Natural disasters? In *Social Nature: Theory, Practice and Politics*, ed. N. Castree and B. Braun. Oxford, UK: Blackwell Publishing, pp. 170–188.

Pelling, M. (ed.) (2003). *Natural Disasters and Development in a Globalizing World*. New York: Routledge.

Rahman, K. M. M., Ensor, J. and Berger, R. (2009). River erosion and flooding in northern Bangladesh. In *Understanding Climate Change Adaptation: Lessons from Community-Based Approaches*, eds. J. Ensor and R. Berger. Rugby, UK: Practical Action Publishing, pp. 39–54.

Ribot, J. C. (2010). Vulnerability does not just fall from the sky: toward multi-scale pro-poor climate policy. In *Social Dimensions of Climate Change: Equity and Vulnerability in a Warming World*, eds. R. Mearns and A. Norton. Washington, DC: World Bank, pp. 47–74.

Schipper, L. (2006). Conceptual history of adaptation in the UNFCCC process. *Review of European Community and International Environmental Law*, 15(1), 82–92.

Schipper, L. (2007). *Climate Change Adaptation and Development: Exploring the Linkages. Tyndall Centre Working Paper Series 107*. East Anglia, UK: Tyndall Centre for Climate Change Research.

Sen, A. (1981). *Poverty and Famines: An Essay on Entitlement and Deprivation*. Oxford, UK: Oxford University Press.

Smit, B., Burton, I., Klein, R. and Wandel, J. (2000). An anatomy of adaptation to climate change and variability. *Climatic Change*, 45(1), 233–251.

Sperling, F. (ed.) (2003). *Poverty and Climate Change: Reducing the Vulnerability of the Poor through Adaptation*. Washington, DC: World Bank.

Williams, G. (2004). Evaluating participatory development: tyranny, power, and (re) politicization. *Third World Quarterly*, 25(3), 557–578.

Wisner, B., Blaikie, P. M., Cannon, T. and Davis, I. (2004). *At Risk: Natural Hazards, People's Vulnerability and Disasters*. London, UK: Routledge.

22

What Next? Climate Change as a Game-Changer
for Policy and Practice

KAREN O'BRIEN AND THOMAS E. DOWNING

There is mounting evidence that climate change influences the frequency, intensity, spatial extent, duration and timing of extreme weather events, which can potentially increase the risk of disasters (IPCC, 2012). Although it is understood that many of the actions that reduce vulnerability to climate variability and extreme events can positively contribute to long-term adaptation to climate change (Sperling and Szekely, 2005; Schipper and Pelling, 2006; Thomalla et al., 2006), the links between disaster risk reduction and climate change adaptation have been weak in practice (UNISDR, 2009). With few exceptions, there is little evidence of the integration of long-term climate change adaptation into current disaster risk management policies. As concluded in the IPCC Special Report on Managing the Risks of Extreme Events and Disasters to Advance Climate Change Adaptation (SREX), 'Closer integration of disaster risk management and climate change adaptation, along with the incorporation of both into local, sub-national, national, and international development policies and practices, could provide benefits at all scales' (IPCC, 2012: 11).

The case studies presented in this book show that responses to past events can provide key insights and lessons for addressing future risks. Collectively, the case studies also illustrate the complex relationship between hazards and vulnerability and point to the importance of understanding risk within dynamic social, economic, political, technological and institutional contexts. However, as discussed by Barnett and colleagues (Chapter 24), there are both barriers and limits to adaptation and few signs that governments are seriously addressing the implications of climate change for long-term decisions on settlement patterns, mobility, infrastructure and the economy.

Climate change introduces many new dimensions to disaster risk management, not only by influencing the physical characteristics of many hazards but also by affecting the capacity to collectively respond to extreme events, uncertainties and surprises (Downing, 2003; Eakin, 2005). Although there is general agreement on the need to integrate disaster risk management and climate change adaptation in policy decisions and practices, there remain many unanswered questions when it comes to long-term adaptation to weather and climate extremes. For example:

- To what extent is historical and contemporary experience with disaster risk reduction transferable to new hazards? The balance between engineered and socio-institutional responses to sea level rise or glacial lake outburst floods, for instance, might not be obvious from lessons learned in managing existing floods (Moser, 2005).
- Which current strategies for preventing or managing disasters might fail in the future as climate change impacts become more and more pronounced? An obvious case is engineered water basins, where much can be done to increase capacities to manage drought or floods, yet where some scenarios of climate change would result in system failures, even with those measures in place (Reeder et al., 2009).
- How does the international finance regime influence planning for disaster risk reduction and climate change adaptation? There is clear agreement that significant adaptation financing is needed for developing countries as part of a global compact on stabilising climate change, but there is no direct analogue for disaster risk reduction (Dellink et al., 2009). In fact, there is a danger that many disaster risk reduction strategies will simply be repackaged as adaptation rather than differentiating the requirements for new and transformative action.

These questions have not yet been adequately addressed in policy circles. Although the experiences from disaster risk management can be considered an essential foundation for long-term adaptation to climate change (Mercer, 2010), we argue that an extrapolation and improvement of current responses is not sufficient. The integration of disaster risk reduction and climate change adaptation into one common policy frame calls for more than 'business as usual', for it draws new variables into the risk equation, including climate change mitigation policies and development policies. Using the metaphor of a game, we discuss the relationship between disaster risk reduction and climate change adaptation, arguing that climate change is a 'game-changer' that calls for new ways of thinking about this relationship. Reflecting on the case studies presented in this book, we then make four observations that challenge current thinking about the integration of disaster risk reduction and climate change adaptation. We conclude that, whether in relation to institutions, policies or practices, effective responses call for more than taking into account the lessons learned and applying them to future scenarios of climate change. Instead, they call for transformative approaches to reducing risk in a changing climate, including not only iterative learning and adaptive management but also addressing climate change mitigation. The next steps in disaster risk reduction and climate change adaptation are more aptly considered to be a 'dance with change' that converges in a more resilient society.

22.1 Playing Together: Disaster Risk Reduction and Climate Change Adaptation

Both in research and in practice, there is a growing move to align the disaster risk reduction and climate change adaptation communities towards the common goals of reducing vulnerability of communities and achieving sustainable development. For example, both research communities recently worked together to produce the IPCC SREX (IPCC, 2012). The collaboration raised numerous issues, for although the two communities share common roots, particularly within the social sciences (e.g. anthropology, geography, human ecology), the language and approaches of the two communities are not identical (see Box 22.1). Indeed, despite a shared goal of reducing vulnerability to climate extremes, disaster risk reduction and climate change adaptation operate as two separate fields of knowledge, each with distinct attributes and communities of practice, governed by different tacit and explicit rules.

Collaboration between the disaster risk and climate change communities proved to be a success in the SREX assessment of the strategies for effectively managing risks and adapting to climate change. In pulling together diverse bodies of literature to assess knowledge on climate change, extreme events, disaster risk and adaptation, the SREX report generated some new perspectives on the changing nature of disaster risk and its implications for development. One conclusion of the report was that 'The interactions among climate change mitigation, adaptation, and disaster risk management may have a major influence on resilient and sustainable pathways' (IPCC, 2012: 20).

While the potential for working together is considered to be great, in reality both the disaster risk and adaptation communities still operate as two teams that are 'playing together', sometimes in competition, sometimes in collaboration, but always as separate teams. Furthermore, the community working towards greenhouse gas mitigation remains peripheral to these efforts, as does the larger development community. Building on the metaphor of a game, we illustrate some of the challenges of integrating policies for disaster risk reduction and climate change adaptation (referred to in this example as DRR and CCA, respectively). We emphasise their interactions at three levels and argue that no matter how much these interactions are improved, the game itself is changing, calling for some new and innovative approaches to building resilience in a changing climate.

In games such as football, basketball and hockey (or, in this case, vulnerability reduction), there are players (NGOs, humanitarian organisations, government institutions, scientists), teams (DRR and CCA communities) and leagues, such as the game organisers that operate at subnational, national or global scales, mostly regarding membership, financing and scheduling (e.g. United Nations International Strategy for Disaster Reduction [UNISDR], United Nations Framework Convention on Climate Change [UNFCCC]). At one level, each player has abilities and makes decisions, including when to shoot or pass the ball (i.e. when, where and how to intervene to

Box 22.1
The language of disasters

Figure 22.1 shows the 100 most common words from the introductions and conclusions of the case study chapters in this book. The case studies are diverse, from Australia to Africa and cover a variety of extremes, from drought to cyclones. But they do not represent a random selection, either of the hazards, the vulnerabilities, or the authors and their perspectives. They represent instead a mainstream view of disasters that offers some important lessons for climate change adaptation. The tag-cloud reveals the strong perspective of natural hazards – dominated by the hazard (flood, storm, fire) and related terms (expected, events, risk, disasters). Quite a bit of the cloud is taken up with tags related to exposure units – the classic language of people at-risk (million, residents) in vulnerable locations (floodplain, basin, river). These are practical case studies, thus the cloud is not dominated by framework terms – exposure and vulnerability are minor, adaptation gets central status but not capacity, mainstreaming, pathway or scenario as might be expected in a discussion of climate change. It is this distinction that is most striking – the language of pathways of climate change adaptation over the coming decades would represent a very different tag-cloud, with words such as learning, leadership, and transformation appearing more prominently.

Figure 22.1 Word cloud from case studies on disaster risk reduction.

reduce vulnerability). At another level, the players interact cooperatively as a team (e.g. through networks and alliances), often in competition with their opponents (e.g. for funding, prestige, media attention, etc.). At the highest level, the leagues establish the boundaries that govern the game (e.g. what types of strategies to prioritise, as defined by the Hyogo Framework for Action or National Adaptation Programmes of Action). At all levels, actions and interactions are guided by rules: An individual player might have tacit rules ('I'm not the best player for infrastructure rehabilitation'), while the team has a number of explicit rules or set plays regarding how to get the ball into play, how to keep the ball, who guards whom and so on (e.g. analytical tools, methods, communities of practice, etc.). Often, a team's game strategy is outlined by the coach (e.g. experts, policy advisors, boundary organisations, etc.). Each team practices, aiming to improve its game.

These strategies and approaches work well at all levels and have facilitated increasing interplay between DRR and CCA. The two communities are becoming better at 'playing together'. However, such an approach is unlikely to succeed over the long run, simply because climate change is transforming the game of vulnerability reduction. Climate change is characterised by systemic changes that result in non-linear impacts, with feedbacks, thresholds and uncertainties. For disaster risk reduction, it is changing the nature of the hazards and influencing social and biophysical vulnerability. For climate change adaptation, it is increasing the risk of disasters associated with climate variability and extreme events. In other words, there is a need for more than combining players, aligning teams and making adjustments or revisions within the leagues. There is a need for a new approach.

22.2 A Spectator's View

There is considerable evidence that the game is changing. In the following sections, we present four observations on the links between disaster risk reduction and climate change adaptation derived from the case studies presented in this book. We specifically consider the lessons in the context of climate change, noting that current policies cannot simply be tweaked to accommodate the need for climate change adaptation. The new game involves proactively creating a resilient and sustainable future by transforming the systems and structures that currently both exacerbate vulnerability and increase the risk of dangerous climate change. It is actually no longer a game but a much deeper engagement with change processes – a dance with change.

Observation 1: Disaster risk reduction currently focuses on building teams to influence the collective behaviour of players in reaching a common goal.

The prevailing assumption in the disaster risk and climate change adaptation communities is that reducing vulnerability to current extreme events is one of the best ways to adapt to a changing climate and that this can be achieved through better teamwork. Consequently, it is considered essential to engage multiple stakeholders and to 'build teams' for coordinated and collective actions. Indeed, the Queensland floods of 2008 emphasised the important role of social capital, strong community networks, community organisations, volunteering and above all the mobilisation of a diversity of people and social groups (King and Apan, Chapter 10). Mason and colleagues (Chapter 9) argue that engagement and motivation of citizens and stakeholders are critical to adaptation policies, which require a strong public and political will to develop.

This sounds simple and straightforward yet can nonetheless be challenging because it often entails engaging with complex planning processes. As Mills and Snook (Chapter 6) show, planning for the future is often difficult, not because of climate uncertainties but because of the rapid and profound changes in the location and composition of at-risk groups. Indeed, the current global economic and financial dynamics add new uncertainties to future plans, which may complicate adaptation as well. Furthermore, both disaster risk reduction and climate change adaptation are often considered to be responses to anomalous situations that are disembedded from development processes. As Eakin and colleagues (Chapter 18: 132) conclude: 'Disaster management is principally considered an issue of social protection and preparedness for temporally specific extreme events, removed from the broader spatial and temporal context of social-ecological development.'

Without discounting the need for effective teams, we recognise that most disasters are (almost by definition) complex social landscapes. There is thus a need to do more than 'pull together the team and respond', but rather to also challenge the embedded structures that create the hazards through transformative responses. Sticking with our metaphor, such transformations are very often beyond the realm of a team and involve instead questioning the game itself.

Observation 2: Despite two decades of focused effort, progress in disaster risk reduction remains highly uneven.

The social toll of disasters around the world is increasing. Despite reductions in mortality, the risk of economic losses associated with tropical cyclones and floods is growing as more economic assets are exposed to such events (UNISRD, 2009; 2011). The disaster losses experienced by low-income households and communities are underreported, as they are often linked to smaller, recurring events (UNISDR, 2009; 2011). The tremendous social and economic impacts of extreme events suggest that disaster risk reduction remains a significant challenge at all scales (e.g. World Bank, 2010).

Coping strategies that reduce disaster risk have long histories: The core strategies are not novel, and many are widely shared as entry points for adaptation. For example, many groups have been adapting to extreme events such as drought over long periods. As Batterbury and Mortimore (Chapter 15) show, the Sahelian people have been adapting to climatic variability and change throughout history, often outside of the public view, often by informal learning and experimentation. More recently, the rapid and widespread adoption of cell-phone technology enables communication among households and facilitates labour management and access to price information (Batterbury and Mortimore, Chapter 15). Such technologies also enable low-cost financial transactions where schemes such as M-Pesa are available (Hughes and Lonie, 2007). Working in the coffee regions of Mesoamerica, Eakin and colleagues (Chapter 18) note that sources of resilience can be found in the farmers' existing ecological knowledge, in their capacities as resource managers and in existing social institutions.

However, local capacity and local knowledge are not always adequately reflected in sectoral plans for disaster risk reduction. There are many examples of inadvertent failures in disaster risk management and efforts to control disaster risk. Kusky (Chapter 4) discusses how flood control measures can actually decrease the carrying capacity of rivers, raising flood levels for any given amount of water. In this case, the strategy of building levees and wing dikes to reduce vulnerability to floods could be ineffective if water volumes increase in the future as a result of climate change. Likewise, reducing vulnerability to wildfires by completely removing vegetation surrounding rural homes is shown by Keeley and colleagues (Chapter 5) to encourage growth of grasses that are sensitive to a substantially longer fire season. This could create new problems, particularly if climate change extends the fire season.

The need for institutional change is well known in disaster risk communities. The challenges for policy are exacerbated when political systems themselves are undergoing change. The 1997 Oder/Odra flood occurred during a period of 'systemic transition from communist regime and centrally planned economy to democracy and market economy' (Kundzewicz, Chapter 13: 94). During this time, the distribution of responsibilities was ambiguous and even conflictual as old laws were abandoned while the new system was not yet firmly established. This sort of socio-institutional change may be the norm on the time scales of climate adaptation.

Despite some very good players here and there, most disaster risk reduction and climate change adaptation teams are fairly weak, with many of them falling below league standards. The leagues themselves are not optimally organised, and financial crises are placing constraints on their flexibility. Instead of focusing on technologies or management strategies aimed uniquely at reducing disaster risk, the task ahead seems to be to move toward a broader view of resilience that includes new institutional approaches.

Observation 3: There are only a few cases in which disaster risk reduction directly contributes to long-term vulnerability reduction, and it is clear that such approaches alone are not sufficient.

The most effective disaster risk reduction approaches involve a process of learning. As Whittaker and colleagues (Chapter 8) note, despite sufficient warnings, many people waited until bush fires directly threatened them before taking action. There is often a lack of social memory, as the Boxing Day 2004 Tsunami (Hettiarachchi and Dias, Chapter 16) reminds us. In this case, public awareness of tsunami risk was very small prior to the event. Nonetheless, the event itself may have contributed to a greater public and institutional awareness and hence better preparedness for future events. However, in the long run, such a 'learning-by-shocking' approach may not be optimal (see Tschakert and Dietrich, 2010). Furthermore, it is not ideally suited to scaling into effective adaptation strategies, particularly in cases where shocks become more frequent and threatening.

High-risk populations (e.g. advanced age, poverty, physical or cognitive limitations) have limited ability to access and act on many risk reduction measures. As demonstrated in the 2003 European heatwave (Pascal et al., Chapter 12), most prevention measures for heat stress are simple yet not easily implemented by the most vulnerable. In East Africa, Goulden and Conway (Chapter 19) emphasise that those who are most vulnerable do not have sufficient access to autonomous adaptation actions that provide livelihood resilience; thus, they need additional protection from the potential damage of floods. These diverse aspects of social capital need to be strengthened at a time when public resources in many countries are diminishing.

Current approaches to disaster risk management are proving to be inadequate in a changing climate. In East Africa, for example, severe floods are less frequent than droughts; thus, inhabitants of lake shore areas are less familiar with flood impacts, which are increasing (Goulding and Conway, Chapter 19). As discussed in the case of the 1997 Odra/Oder flood (Kundzewicz, Chapter 13), climate change is likely to increase the number of disasters and possibly their magnitude. Nonetheless, climate change is about more than just drought, and although many will be able to draw on local knowledge and social learning, there may be limits to adaptation to complex changes in social-ecological systems.

Observation 4: Climate change adaptation is a development pathway that extends beyond reducing current weather-related impacts.

Many of the case studies in the disaster risk reduction literature call for a change in perspective that acknowledges risk as an inevitable component of lives and livelihoods. One of the obvious differences between studies of disaster risk reduction and climate adaptation is the longer-term vision of climate change futures. Bridging between the imperative of reducing current vulnerability and reducing the impacts of climate change over the next few decades to a century-long time scale is the essential challenge of climate change adaptation. Since predicting impacts on these time scales is impossible, adaptation is increasingly seen as a process of social and institutional change, management and learning (Moser and Ekstrom, 2010).

Botterill and Dovers (Chapter 7) note that climate change adaptation can be addressed by 'normalising' adaptation – that is, by reform across many sectors to anticipate increased climate variability. They trace the long and variable process of policy reform in dealing with drought in the Murray-Darling Basin of Australia. In this case, the discourse shifted as droughts went from being considered as an exception to being expected. They also note the potential for policy reversal, particularly as value conflicts arise.

As demonstrated in the case of cyclone Tracy (Mason et al., Chapter 9), there is an enormous pressure to return to 'normal' and get back to 'business as usual', and delays or changes are quickly overruled following most large disasters. This poses a challenge for climate change adaptation. If the periods between disasters can be considered a 'time out' for reflection, then it may be possible to take into account lessons, systems perspectives and so on. However, as in the case of insurance for windstorms in Europe (Ulbrich et al., Chapter 11), a temporary absence of extreme damage led to a decrease in the price of storm reinsurance parallel to the surplus in reinsurance capacity after major events. From a long-term perspective, they argue, premiums and reinsurance products may be insufficient, and there is a need to adapt these to new situations rather than to base them only on the most recently experienced losses. The need to adapt to new situations is important. As Kundzewicz and colleagues (Chapter 20: 146) emphasise, although it is recognised that intense precipitation can increase in a warming climate, 'reliable and detailed quantification of flood statistics is very difficult to obtain for the past to present and is virtually impossible to obtain for the future'. The predict-and-provide model of dealing with disasters is not tenable for the future.

Vellinga and Aerts (Chapter 14) show that the surprising 1953 tidal surge in the North Sea should not have been a surprise given that insufficient funds had been allocated for maintenance. In fact, prudent plans to reduce vulnerability had been deemed too ambitious and costly – what might be deemed a failure of planning. However, this is such a dominant feature of disaster studies that we must question whether the nature of the challenge has been adequately captured. A larger frame of reference – usually labelled as a systems approach – is called upon by Mason and colleagues (Chapter 9). Applying lessons to isolated components of a system can backfire, as failure can simply occur at the next step in the structural chain. These authors advocate consideration of the structural system as a whole, including a suitable level of redundancy.

Viewing climate change adaptation from a systems perspective calls for much more than managing disaster risk. It involves looking at the larger context of change processes and considering how different development pathways influence the social, economic, political and institutional context in which risk is both created and experienced.

22.3 A Dance with Change

What comes next in terms of both policy and practice? This is not an easy question to answer, as responses involve all levels of government, NGOs, the private sector, development agencies and civil society. Furthermore, responses at all levels and sectors are likely to include a combination of adjustments to current policies and

Climate change adaptation

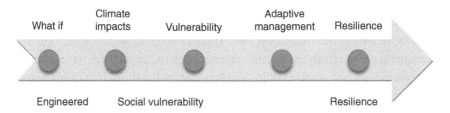

Figure 22.2 Progression of major approaches in disaster risk reduction and climate change adaptation.

actions and transformative changes (O'Brien et al., 2012). What seems clear is that more coherent, creative and flexible strategies are needed to bring together the disaster risk reduction and climate change adaptation communities for managing new and in some cases unprecedented risks.

The 'game' that we have discussed above is changing as a result of climate change, and the result may be considered something more akin to a dance with change. This new dance acknowledges (1) the nonlinear implications of climate change for disaster risk reduction and (2) relationships between development policies and climate change adaptation. The steps for this dance have not yet been created, for they will depend on the relationship between actions to mitigate climate change and address social vulnerability. Moving from a 'game' to a 'dance' recognises that strategies and approaches that worked in the past may not be appropriate to the future and that new approaches to learning and leadership are necessary. The dance represents a creative engagement with the future, recognising that humans are responsible for many of the changes that contribute to growing risks and vulnerability – and that they are responsible for creating alternative pathways.

Long-term adaptation involves both incremental adjustments and transformational changes in technologies, financial structures and ideas about disaster risk. Over the past few decades, both the disaster risk reduction and climate change adaptation paradigms have been shifting (see Figure 22.2). Early framing posed adaptation in terms of 'what if' scenarios: If climate changes as X, what would be the consequences for Y? This naturally led to a 'predict-and-provide' approach based on scenarios of climate impacts. The equivalent history for disaster risk reduction is the longstanding tradition of engineering solutions that 'protect' vulnerable areas. Equally longstanding in the hazards community, however, has been approaches based on vulnerability as socially produced conditions of hazard (Wisner, 2004). This has enabled a wider perspective on actions to reduce risk, including changing access to resources, creation of social safety nets and recognition of local capacity and indigenous knowledge. The vulnerability-first approach to climate change adaptation was proposed two decades ago (Downing,

1991) but only gathered momentum in the IPCC Fourth Assessment Report (Adger et al., 2007). Both disaster risk and climate change adaptation communities are gradually moving beyond static representations of current vulnerability to managing dynamic change as pathways of adaptive learning and resilience.

A dance with change involves continuous learning, not as in 'learning the rules of the game' but as an iterative process of joint problem solving and reflection – a process that builds resilience in the present while developing the capacity to shape the future (Armitage et al., 2008; Moser, 2010). The learning loop framework (Argyris and Schön, 1978) is also highly relevant to a dance with change, as it divides learning into three loops that reflect different approaches to change, similar to acquiring different sets of dance skills. With single-loop learning, changes are made based on the difference between observations and expectations; it is about the practical actions that are appropriate to identified objectives in disaster risk reduction. Double-loop learning questions whether current objectives and goals are appropriate, and corrective actions are then taken. Triple-loop learning questions deeper assumptions, including social structures, norms and values. These different approaches to learning play important roles in disaster risk reduction and climate change adaptation (Lavell et al., 2012; O'Brien et al., 2012). However, triple-loop learning represents perhaps the most creative dance style and the one that is most likely to result in transformative responses to climate extremes.

Adaptive management reflects an incremental and iterative learning-by-doing process that takes into account changes in external factors, such as changing climate extremes (O'Brien et al., 2012). One of the consequences of a shift from a predict-and-provide approach to adaptive management is to refocus adaptation on actual institutional practices. Lessons learned in screening development portfolios show that an approach based on building capacity rather than understanding all possible risks is an essential starting point. For example, the Global Climate Adaptation Partnership (2011) has developed a system for the African Development Bank based on the Bank's project cycle (see Figure 22.3). At each stage of project development, the Climate Safeguards System comprises practical procedures. The initial screening of each project into one of three categories identifies the level of climate adaptation planning that is required and the supporting level of due diligence in meeting the Bank's Climate Risk Management Strategy. Embedding practical tools in institutions such as the African Development Bank can be considered an initial step in building effective capacity for 'adapting well' into the future.

22.4 Conclusions

We have argued in this chapter that current approaches to disaster risk management are likely to be insufficient for long-term adaptation. Why? Because climate change is about more than changing variability and extremes – it is also about systemic changes

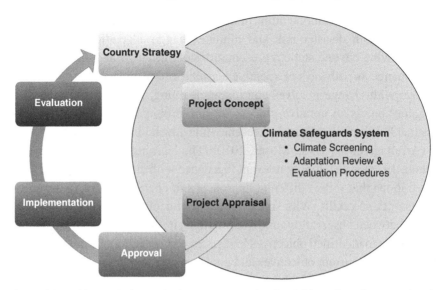

Figure 22.3 Climate Safeguards System (CSS) in the African Development Bank. Projects are screened at the design stage into one of three classes for further consideration in a module on Adaptation Review and Evaluation Procedures. The CSS supports development of country strategies through an information base and project appraisal in a checklist of requirements (modified from African Development Bank Group, 2011).

to the atmosphere, biosphere, cryosphere and hydrosphere and their interactions with social, economic, cultural, political, technological and institutional changes. This means that we can no longer conceptualise hazards as 'special events', but instead we need to think of them as part of a new set of global dynamics. Rather than influencing only the outcomes, human activities are influencing the systems themselves. Using the analogy of a sports game, we have argued that it may no longer be sufficient to improve capacities, hone skills or manage strategies when the game itself is changing. The new game is a dance with change, and it requires new skills and new steps to move forward on a path to sustainability.

References

Adger, W. N., Agrawala, S., Mirza, M. M. Q. et al. (2007). Assessment of adaptation practices, options, constraints and capacity. In *Climate Change 2007: Impacts, Adaptation and Vulnerability. Contribution of Working Group II to the Fourth Assessment Report of the Intergovernmental Panel on Climate Change*, eds. M. L. Parry, O. F. Canziani, J. P. Palutikof, P. J. van der Linden and C. E. Hanson. Cambridge, UK: Cambridge University Press, pp. 717–743.

African Development Bank Group (2011). *Climate Safeguards System (CCS). Climate Screening and Adaptation Review and Evaluation Procedures Booklet.* Tunis-Belvedère, Tunisia: African Development Bank Group. Accessed 12 December 2012 from: <http://issuu.com/mo_c/docs/afdb_css_basics_en>.

Argyris, C. and Schön, D. (1978). *Organizational Learning: A Theory of Action Perspective.* Boston: Addison-Wesley.

Armitage, D., Marschke, M. and Plummer, R. (2008). Adaptive co-management and the paradox of learning. *Global Environmental Change*, 18, 86–98.

Dellink, R., den Elzen, M., Aiking, H. et al. (2009). Sharing the burden of financing adaptation to climate change. *Global Environmental Change*, 19(4), 411–421.

Downing, T. E. (1991). Vulnerability to hunger and coping with climate change in Africa. *Global Environmental Change*, 1(5), 365–380.

Downing, T. E. (2003). Lessons from famine early warning systems and food security for understanding adaptation to climate change: toward a vulnerability adaptation science? In *Climate Change, Adaptive Capacity and Development*, eds. J. B. Smith, R. J. T. Klein and S. Huq. London, UK: Imperial College Press, pp. 71–100.

Eakin, H. (2005). Institutional change, climate risk, and rural vulnerability: cases from central Mexico. *World Development*, 33, 1923–1938.

Global Climate Adaptation Partnership (2011). *Adaptation Space: Switching Paradigms: The Landscape of Adaptation*. Oxford, UK: Global Climate Adaptation Partnership.

Hughes, N. and Lonie, S. (2007). M-PESA: Mobile money for the "unbanked" turning cellphones into 24-hour tellers in Kenya. *Innovations*, winter and spring. Accessed 12 December 2012 from: <http://www.mitpressjournals.org/doi/pdf/10.1162/itgg.2007.2.1-2.63>.

IPCC – Intergovernmental Panel on Climate Change (2012). Summary for policymakers. In *Managing the Risks of Extreme Events and Disasters to Advance Climate Change Adaptation. A Special Report of Working Groups I and II of the Intergovernmental Panel on Climate Change*, eds. C. B. Field, V. Barros, T. F. Stocker et al. Cambridge, UK: Cambridge University Press, pp. 3–21.

Lavell, A., Oppenheimer, M., Diop, C. et al. (2012). Climate change: new dimensions in disaster risk, exposure, vulnerability, and resilience. In *Managing the Risks of Extreme Events and Disasters to Advance Climate Change Adaptation. A Special Report of Working Groups I and II of the Intergovernmental Panel on Climate Change*, eds. C. B. Field, V. Barros, T. F. Stocker et al. Cambridge, UK: Cambridge University Press, pp. 25–64.

Mercer, J. (2010). Disaster risk reduction or climate change adaptation: are we reinventing the wheel? *Journal of International Development*, 22, 247–264.

Moser, S. C. (2005). Impact assessments and policy responses to sea-level rise in three US states: an exploration of human-dimension uncertainties. *Global Environmental Change*, 15, 353–369.

Moser, S. C. (2010). Now more than ever: the need for more socially relevant research on vulnerability and adaptation to climate change. *Applied Geography*, 30(4), 464–474.

Moser, S.C. and Ekstrom, J.A. (2010). A framework to diagnose barriers to climate change adaptation. *Proceedings of the National Academy of Sciences of the United States of America*, 107(51), 22026–22031.

O'Brien, K., Pelling, M., Patwardhan, A. et al. (2012). Toward a sustainable and resilient future. In *Managing the Risks of Extreme Events and Disasters to Advance Climate Change Adaptation. A Special Report of Working Groups I and II of the Intergovernmental Panel on Climate Change*, eds. C. B. Field, V. Barros, T. F. Stocker et al. Cambridge University Press: Cambridge, UK, pp. 437–486.

Reeder, T, Wicks, J., Lovell, L. and Tarrant, O. (2009). Protecting London from tidal flooding: limits to engineering adaptation. In *Adapting to Climate Change: Thresholds, Values, Governance*, eds. W. N. Adger, I. Lorenzoni and K. L. O'Brien. Cambridge, UK: Cambridge University Press, pp. 54–63.

Schipper, L. and Pelling, M. (2006). Disaster risk, climate change and international development: scope for, and challenges to, integration. *Disasters*, 30(1), 19–38.

Sperling, F. and Szekely, F. (2005). *Disaster Risk Management in a Changing Climate. Informal Discussion Paper Prepared for the World Conference on Disaster Reduction on Behalf of the Vulnerability and Adaptation Resource Group*. Washington, DC: Vulnerability and Adaptation Resource Group.

Thomalla, F., Downing, T., Spanger-Siegfried, E., Han, G. and Rockström, J. (2006). Reducing hazard vulnerability: towards a common approach between disaster risk reduction and climate adaptation. *Disasters*, 30(1), 39–48.

Tschakert, P. and Dietrich, K. (2010). Anticipatory learning for climate change adaptation and resilience. *Ecology and Society*, 15(2), 11.

UNISDR – United Nations International Strategy for Disaster Reduction (2009). *Global Assessment Report on Disaster Risk Reduction: Invest Today for a Safer Tomorrow*. Geneva, Switzerland: United Nations International Strategy for Disaster Reduction.

UNISDR – United Nations International Strategy for Disaster Reduction (2011). *2011 Global Assessment Report on Disaster Risk Reduction: Revealing Risk, Redefining Development. Geneva*, Switzerland: United Nations International Strategy for Disaster Reduction.

Wisner, B. (2004). Assessment of capability and vulnerability. In *Vulnerability: Disasters, Development and People*, eds. G. Bankoff, G. Frerks and T. Hilhorst. London, UK: Earthscan, pp. 183–193.

World Bank (2010). *Natural Hazards, UnNatural Disasters*. Washington, DC: World Bank.

23

Barriers and Limits to Adaptation: Cautionary Notes

JON BARNETT, COLETTE MORTREUX, AND W. NEIL ADGER

In this chapter, we provide a synthesis of the empirical chapters of this collection by examining what they reveal about the limits and barriers to climate change adaptation and the risk of maladaptation. The concept of adaptation is therefore central to this chapter. While there are many definitions and typologies of adaptation (e.g. Smit et al., 2000), adaptation here is understood as the process of adjusting to climate change. Adaptation takes one or more of four forms: Reducing the exposure of entities at risk; reducing the sensitivity of entities at risk; increasing the capacity of entities at risk to avoid risks; and taking advantage of new opportunities created by a changing climate.

Research and policy on adaptation have evolved, slowly, away from conceptual explanations, typologies, frameworks and wish-lists to include more serious consideration of the challenges of implementing adaptation given existing political and policy environments. In this vein, Adger and Barnett (2009) identify four reasons for thinking that adaptation may not happen as easily or effectively as initially envisaged. First, they argue that the scale of the change in climate, the interconnectedness and cascading nature of potential impacts and the possibility of significant surprises may be such that the time available for adaptation to take effect is less than previously imagined. The second reason for concern is that no amount of adaptive capacity – that is, potential for adaptation – ensures that adaptation will actually happen. There are, in effect, barriers that impede adaptation from taking place. Third, they suggest that adaptation actions that have been taken thus far indicate that adaptation decisions can be ineffective, if not actually make things worse, at least for some groups within and between generations. Finally, they argue that there are limits to what adaptation can achieve in the sense that there can be impacts of climate change (or of climate change policies) that involve irreversible losses of things individuals care about. In short, Adger and Barnett (2009) are not sanguine about adaptation given barriers that impede best practices, limits to what can be achieved by purposeful actions and adverse consequences that increase risks. It is these barriers, limits and maladaptations that are the focus of this chapter.

This book contains eighteen cases of disasters. Disasters are extremely useful analogues for studying adaption to climate change, if for no other reason than because climate change will most likely bring with it more intense extreme events of the kind discussed in this book; firestorms, hurricanes, droughts, floods and heatwaves are all expected to increase in intensity, if not frequency, under almost all climate scenarios (IPCC, 2007). Thus, disaster risk management is a key subset of climate change adaptation (Schipper and Pelling, 2006).

Moreover, the distributional consequences of responses to disasters – that is, the ways the costs and benefits of responses to disasters are spread across present and future generations – also tells us much about the likely outcomes of adaptation. Decisions are made (or not) about how to reduce the risk of future extreme events, and this is much the same as with adaptation, which is a process of making decisions to reduce risks arising from climate change. In both cases, then, who wins and loses is a function of various influences, such as the kinds of information that inform decisions, the identification of risks to be avoided, the policy community that engages in decision making and the political economy of public policy. As a consequence, in both cases, for example, some people benefit from responses and some do not; some people's risks are reduced while others' are not and some may even be amplified; and trade-offs are made, for example, between ecosystem goods and services and public safety, the rights of present and future generations and between some groups and others.

This chapter first presents some theory to differentiate between the concepts of barriers to adaptation, limits to adaptation and maladaptation. It then synthesises and draws insight from the cases studies in this collection organised according to each of those themes. Conclusions are then drawn.

23.1 Barriers, Limits and Maladaptation in Theory

Barriers to Adaptation

Barriers to adaptation are 'obstacles that can be overcome with concerted effort' (Moser and Eckstrom, 2010: 22027). These explain the difference between the widespread acceptance of the need for adaptation, the many plans and processes that are in place and the largely incomplete and often conflicted outcomes observed thus far. There is, however, a grey area between barriers (impediments that should be possible to manage) and limits to adaptation (thresholds at which adaptation fails to avoid a climate impact). This is often a matter of experience and judgment. For example, relocating a coastal village at the cost of $10 million might seem to the residents of that village to be a barrier in that they expect it will be difficult but not impossible to leverage the resources. The high cost may prove ultimately to be a limit to adaptation if no individuals or public bodies are willing to pay (and therefore adaptation fails to avoid the impacts of sea-level rise on the village).

Given the complex cross-sectoral, multi-scale and temporal nature of climate change, there may be more barriers to adaptation policy than other sectors in which policy is made. The list of barriers is therefore potentially large. They can include cultural (defined here as shared symbols, beliefs and practices) barriers, including the values and identities of communities at risk; routine social practices, including consumption, movement, social interaction, health seeking and recreation; public acceptance of climate change and the need to adapt; the use of public services and goods; feelings of empowerment; forms of civil society organisation perceptions of risk; and trust in and attitudes towards science and decision makers (see, for example, Ellemor, 2005; Grothmann and Patt, 2005; Heyd, 2008; Kuruppu, 2009; Swim et al., 2009; Leiserowitz et al., 2010; Neilsen and Reenberg, 2010; Adger et al., 2011). This is not to say these factors are necessarily barriers – they can indeed be enablers of adaptation, too – but it is to say that adaptation is likely to be impeded and slow in communities that, for example, do not believe in climate change, do not feel that they can take actions to adapt or do not trust decision makers.

There are also institutional (defined here as underlying structures, norms and processes governing decision making) barriers (Dovers, 2005). These can include how the power in decision making is distributed across society; how the goals of adaptation are defined; the knowledge that informs decision making; perceptions of the vulnerability of other groups and sectors; the distribution of risks and responsibilities across the public and private sectors; inter-jurisdictional coordination within governments; the degree to which different levels of government work together; leadership; the forms and nature of communication within and between groups; the timeframes for adaptation processes; statutory and common law; property rights; the ability of organisations to learn; and the availability and use of resources (see, for example, Adger, 2003; Moser, 2005; Næss et al., 2005; Few et al., 2007; Lorenzoni et al., 2007; McDonald, 2007; Burch, 2010; Dovers and Hezri, 2010; Storbjörk, 2010; McDonald, 2011). Again, these factors can enable or constrain adaptation, but, as with the cultural factors, they are at least in theory mutable in as much as they are socially constructed and so can be reconstructed.

Finally, and not insignificantly, financial resources can be a significant barrier to adaptation. Whilst it is premature to cost global adaptation, as adaptation strategies are mostly undecided and their associated benefits and costs remain uncertain, it is safe to say that adaptation is likely to be very expensive, with some estimating it to cost several USD billion per year (Agrawala and Fankhauser, 2008). Meeting financial costs can present significant barriers to adaptation (see, for example, Mendelsohn, 2000; Farber, 2007). Considering the current global financial climate and how this is manifesting itself on smaller scales, sourcing the funds to pay for adaptation is likely to present a further barrier (if not a limit) to adaptation.

Limits to Adaptation

Unlike barriers, which are mutable impediments to adaptation in the sense that they can be overcome through a reasonable effort (Burch, 2010), limits to adaptation are the points at which adaptation fails to avoid climate change impacts (Adger et al., 2009). There are thresholds beyond which adaptation cannot avoid ecological impacts. For example, there seems likely to be a threshold beyond which no amount of human action can avoid repeated and severe coral bleaching, with subsequent impacts on species diversity and function (Donner et al., 2005). There are also social limits: For example, no amount of adaptation can avoid damages to cultures that will be incurred when coastal lands are submerged by sea-level rise or when people relocate from lost homelands (Barnett and Adger, 2003).

Limits to adaptation arise from economic, technical or social valuation processes. First, some adaptations may be technically feasible but simply too expensive. For example, the costs of protecting cities from sea-level rise are less than the costs of the impacts (e.g. Bigano et al., 2008) and so are likely to be paid. However, the same may not be said for protecting rural coastal settlements. Second, there are thresholds beyond which available technologies cannot avoid climate impacts. For example, under certain climatic conditions, snow-making machines may be able to sustain snow cover for the purposes of skiing (if not for species dependent on the snow-pack), but ultimately climate may change to the point at which snow making is no longer possible (Pickering and Buckley, 2010). Similarly, there may be limits to engineering solutions to avoid flooding in certain places under extreme scenarios of change (Reeder et al., 2009).

The third reason there are limits to adaptation is because social groups may judge adaptation actions to have failed. Because diverse groups value things differently, what may be perceived as a successful adaptive response from one point of view may not be perceived the same way by others. For example, relocating populations from islands at risk of sea-level rise is perceived by some as a feasible solution (e.g. Kelman, 2008), but research from islands at risk reveals that those at risk not do not see this as an adaptation but as an impact of climate change transmitted through adaptation policy (Mortreux and Barnett, 2009; Kuruppu and Liverman, 2011). Thus, the limits to adaptation are, to some extent, in the eye of the beholder.

Maladaptation

Maladaptations are actions taken ostensibly to avoid or reduce vulnerability to climate change that impact adversely on or increase the vulnerability of other systems, sectors or social groups (Burton, 1997; Barnett and O'Neill, 2010). Barnett and O'Neill (2010) identify five pathways to maladaptation, which they demonstrate using the example of a desalination plant in Melbourne, Australia. First, they suggest that if an adaptation action leads to significant increases in greenhouse gas emissions, this is maladaptive in that it creates a positive feedback that amplifies changes in climate. Second, they suggest that adaptation actions that disproportionately burden the most vulnerable are maladaptive in that they increase risk for those most at risk. Third, they argue that adaptations that have high opportunity costs are maladaptive in that they come at the cost of more numerous adaptations. Fourth, they argue that an adaptation action is maladaptive if it creates a moral hazard in that it reduces the incentives for actors to adapt and defers responsible action. Finally, they argue that adaptations that 'lock in' solutions over many decades (these are typically engineering-type responses) are maladaptive by reducing the portfolio of potential options available in the future.

23.2 Barriers, Limits and Maladaptation in Practice

Barriers to Adaptation

The chapters in this collection reveal that there are considerable barriers to disaster management. Andrew Garcia's chapter on hurricane Katrina shows that a strong commitment to engineering solutions created barriers to effective disaster management in New Orleans (Chapter 3). He shows that the hurricane protection system (HPS) designed to protect the city was never completed because of a lack of funding, delays because of design problems, legal and political challenges for environmental concerns and dated risk-estimation methodologies. Thus, at the time of hurricane Katrina, the HPS was incomplete, increasing the risk to the city given misplaced confidence in its effectiveness and poor communication of the risk the hurricane posed to the residents of New Orleans.

The problem of cognitive barriers to adaptation arises in the chapter by Whittaker, Handmer and Karoly on the 2009 Victorian bushfires (Chapter 8). During the 2009 fires, the 'stay and defend or leave early' policy was not particularly effective, as some people misjudged the risk, preferring to 'wait and see' on the day, leading to last-minute departures and a higher rate of death. Despite a subsequent Royal Commission and changes in fire management policies, including a new warning system, the experience of the 2009/10 bushfire season showed that people were still reluctant to leave their homes at the appropriate time. That people are still reluctant to leave despite repeated attempts to make this a response to risk suggests that this may not be a cognitive barrier (in that it may at some stage be overcome) but rather is simply a limit to what adaptation can be achieved in this context.

Chapters 4 (Kusky), 8 (Whittaker et al.), 13 (Kundezewicz) and 16 (Hettiarachchi and Dias) each in their own way identify the constraints that land use places on adaptation. In each case, pressures to develop new lands – in these cases at risk of flooding and fire – act as barriers to effective disaster risk management. The underlying problems here are of inadequate recognition of and warning about risk, the short-term exposure of developers to risk and the way in which purchasers inherit risk and weaknesses in statutory planning to prevent properties in hazardous locations from being developed. In their discussion of recovery from the 2004 tsunami in Sri Lanka, Hettiarachchi and Dias show that shortages of land slowed the reconstruction process (as did shortages of labour and materials). Importantly, they too point to development pressures creating barriers to effective adaptation, as initially declared 'no-build' zones subsequently became settled because of demographic, political and economic pressures. As in the case of the central European flooding, economic and political transitions can also create barriers to effective risk management.

Various chapters show that financial resources impede adaptation. An inability to meet costs was a reason the HPS in New Orleans was not complete and was the cause of the delay in rebuilding the dykes that were damaged during the Second World War in the Netherlands. Chapter 16 shows that the reconstruction of houses after the tsunami in Sri Lanka was delayed by the cost of materials (Hettiarachchi and Dias), and Chapter 18 shows that the delay in recovering from tropical storm Agatha in Mexico was because of the costs of reconstruction (Eakin et al.). Chapter 17 argues that the costs of recovering from cyclone Sidr in Bangladesh have been so prohibitive that reconstruction was still incomplete four years later (Paul and Rahman). Poverty was a considerable cause of mortality in the 1970s drought in northern Nigeria, it impeded the ability of smallholder coffee farmers in Guatemala and Chiapas to adapt to disasters and it was the cause of the relatively greater vulnerability of households flooding in the upper White Nile (Goulden and Conway, Chapter 19).

Communication of risk is a key barrier identified in many chapters. For example, in New Orleans, poor communication reduced the number of people who evacuated (Garcia, Chapter 3), and in Chapter 5, it was argued that the risks of wildfire and the risks of wildfire management strategies have been poorly communicated to households (Keeley et al.). In the case of extreme heat in Kansas City, there remain challenges in public education on the risks of extreme heat events. In their examination of water policy in the Murray-Darling Basin in Chapter 7, Botterill and Dovers also point to the failure to properly consult stakeholders as a reason for the troubled history of water management in the basin. The case of the Victorian bushfires in 2009, discussed in Chapter 8, also identified problems with the communication of risk in order to engage and motivate people (Whittaker et al.).

Limits to Adaptation

A number of chapters speak to the technological limits to adaptation. The problems with continued reliance on embankments to prevent flooding in the Yangtze, for example, are explained in Chapter 20 (Kundezewicz et al.). Chapter 3 shows that while barriers to implementation reduced the effectiveness of the HPS in New Orleans, it is also the case that the magnitude of hurricane Katrina overwhelmed the system (Garcia). The case suggests that there may always be limits to the protection afforded by engineering defences given that they failed in even a wealthy and technologically advanced society such as the United States (similarly, the failure of coastal defence systems to protect the north-east coast of Japan from the tsunami in March 2011). Similarly, in the case of flood control in the Mississippi River and in central Europe, discussed in Chapters 4 (Kusky) and 13 (Kundezewicz), it is clear that there are limits to what technologies can achieve with respect to flood defence.

In the case of the floods in Charleville and Mackay in 2008, flood levees also failed to prevent flooding because of intense rainfall within the levees (King et al., Chapter 10). King and colleagues suggest that the future of Charleville is under question should climate change bring about repeated floods of the magnitude of 2008, despite a high degree of resilience centred on extensive social capital. They suggest that relocation may be necessary, but that this ironically might impact on the social networks and bonds that people value and that served them during the 2008 floods. This raises the possibility that the residents of Charleville might not consider relocation to be a desirable adaptation to climate change and may be resistant, as was the case with efforts to resettle flood-affected people in Uganda (Goulden and Conway, Chapter 19).

The social impacts of population displacement as an adaptation strategy are also raised in the case of the evacuation of 35,000 people from Darwin after cyclone Tracey left only 6 per cent of houses in a habitable state (Mason et al., Chapter 9). Residents were away from their homes for many months, and this long evacuation period is said to have caused significant psychological damage to many residents at the time. However, the case is one of an effective adaptation in the longer term, as the delay in return allowed sufficient time to re-evaluate and reform building design standards, which has lead to a reduction of risk from cyclones in Darwin and Australia more broadly.

The effects of mobility on adaptation are also discussed in the chapter on the Sahel by Batterbury and Mortimore (Chapter 15). In contrast to the problems associated with large-scale displacement after cyclone Tracy and the risk of relocation of Charleville, Batterbury and Mortimore suggest that movements that are more voluntary have fewer social costs and greater benefits to households. In this sense, such movements seem likely to be judged by people in northern Nigeria to be effective adaptations to drought.

In their chapter on insurance against windstorm damage, Ulbrich, Leckebusch and Donat (Chapter 11) raise the spectre of financial limits to adaptation. They suggest that problems in establishing scenarios of future windstorms mean that insurance companies are struggling to set appropriate rates for their customers, with the risk that premiums will not be high enough to cover loss repayments. They show that changes in the ways roofs are constructed reduce losses, but still, the problem of determining future risks raises the possibility of under-insurance against disasters.

Across the chapters there is warning about the magnitude of changes and the potential for surprise surpassing the ability of social systems to cope. Hurricane Katrina, the Victorian Bushfires, the Chinese and Queensland and central European floods, the Boxing Day tsunami and the 2003 heatwave in France are all examples of extreme events that caught society unprepared. In some cases, responses have been made that seem likely to have increased preparedness and reduced losses to future events of a significant magnitude. For example, the new national prevention strategy in France has statutory provisions that seem likely to have led to fewer deaths during a subsequent, albeit more moderate, heatwave (Pascal et al., Chapter 12). Nevertheless, the larger lesson is clear enough: Environmental perturbations can be large and surprising and surpass the limits of society to cope. That the intensity of such events seems likely to increase because of climate change suggests that there is a need to more deliberately plan for surprises and imagine worst-case scenarios in order to minimise the risk that coping thresholds will be crossed more frequently in the future.

Maladaptation

Andrew Garcia's chapter points to the maladaptive effects of engineering solutions, arguing that they create a path-dependence in hurricane defences that, once pursued, is hard to change. In his chapter on flood control along the Mississippi River, Timothy Kusky (Chapter 4) makes a similar argument – that once decision makers become committed to levee banks as the primary means of flood protection, alternative strategies are devalued. The problem of sunk costs has meant that there has been increasing investment in levee banks, even though it has long been realised these do not work and may indeed increase flood risk. The case of flood control in the Mississippi shows that flood managers are not adapting to changing flood risks in that there has been no response to new knowledge.

In suggesting that systems of risk communication, population control and evacuation may be more effective means of flood defence, both the New Orleans and Mississippi cases suggest that engineering solutions have high opportunity costs as well as high carbon costs (also a problem in the increased use of air conditioners to manage heat stress in Kansas City – see Mills and Snook, Chapter 6).

In his discussion of what is perhaps the world's largest and most effective flood defence system, Pier Vellinga and Jeroen Aerts show that while Deltaworks has effectively reduced flood risk, it has led to adverse environmental effects (Chapter 14). These include the salination of freshwater bodies, changes in sediment dynamics and impacts on the habitats of migratory birds.

Flood defences have also reduced incentives for households to take appropriate measures to prepare for flooding in New Orleans. This problem of moral hazards is also present in drought policy in Australia, as discussed by Botterill and Dovers in Chapter 7, who suggest that drought relief paid on the basis of drought being an 'exceptional' event have sustained unviable farming businesses. In their chapter, Mason, Haynes and Walker (Chapter 9) also point to the false confidence people had in the ability of their homes to withstand damages from cyclones as a reason for inaction in advance of cyclone Tracy. These flooding cases suggest that engineering systems to resist floods can actually increase flood risk through path dependence, opportunity costs and reduced incentives for risk management at smaller scales.

The case of wildfires in southern California in Chapter 5 shows that management strategies can also have the unanticipated effect of increasing fire risk (Keeley et al.). Laws introduced in southern California to reduce fire risk require homeowners to reduce fuel loads for a zone of 30 metres around their homes. Homeowner enthusiasm to reduce fire risk has lead to over-zealous clearing, with the larger cleared areas having adverse effects on wildlife habitats and soil moisture, and they have ultimately increased fire risk.

23.3 Conclusions

Adaptation is an imperfect process that is prone to barriers to best practice, limits to what can be achieved and uneven consequences for different groups across space and time. One of the highest priorities for adaptation for governments seeking to protect their citizens is to reduce the likelihood of disaster. Governments in particular seek to reduce those elements of disaster for which they have responsibility or can be blamed. While many governments have long experience of disaster risk reduction, as discussed in every chapter in the book, few are grappling seriously with the longer-term decisions on settlement patterns, mobility, infrastructure and the economy that constitute adaptation to climate change.

For both adaptation and disaster risk reduction, outcomes of actions are contingent upon many factors, including: Which risks are identified and prioritised and which are ignored; who is judged to be at risk; who has influence in making decisions; which information is used to make decisions; the costs of actions and who pays them; who is responsible for implementing decisions; the instruments used to implement decisions; and how much time there is to implement actions (and how much time it takes).

In this chapter, reflecting on the volume as a whole, we have focused on the outcome of disaster and the distribution of the related risks and consequences for different parts of society. One could judge adaptation to be successful if risk is reduced, if risk for the most vulnerable groups is reduced in particular and if significant costs are avoided. But it is worth noting that there are other dimensions to success. Adaptation is also, as we have discussed, a process. Success in adaptation can therefore also be judged by whether that process is legitimate and inclusive as well as sustainable over the long run. Adaptation is ultimately unsustainable where 'solutions' to weather-related disasters are imposed rather than following a dynamic process of community consultation.

A further dimension of success would involve upholding a universal right not to be exposed to a harmful or unsafe environment. Such a rights-based approach to disaster equally focuses on the most vulnerable and the circumstances that force them to live in hazardous places, to live in unsafe housing conditions or to be exposed to risk through their livelihoods. The rights approach, of course, raises a significant further dimension in the context of climate change: Anthropogenic climate change is an imposed harm and hence disasters associated with climate change (as opposed to natural climate variability) represent an imposed harm. Singer (2011), Gardiner (2011), Caney (2010), and many others make the case that such actions, if attributable to human action, cannot be justified by the consumption benefits of the energy use that ultimately impose the harm. The notion of rights not to be exposed to disaster puts the cases discussed in this volume into a whole new moral and legal territory if they can be attributed to knowable and avoidable human actions.

Some chapters in this book give cause for optimism. In cases such as the French heatwaves, cyclone Tracy, floods in Central Europe, storm surges in the Netherlands, drought in the Sahel and tsunami in Sri Lanka, lessons have been learned and changes have been made that seem to have reduced vulnerability to future hazards. In other cases, such as hurricanes in New Orleans; flooding in the Mississippi, north Queensland, the upper Nile and the Yangzi; fires in California and Victoria; drought in the Murray-Darling Basin, and cyclones in Bangladesh, there appears to have been little learning such that vulnerability has been reduced.

As analogues for adaptation, the cases reveal that wealth and technological capacity do not necessarily determine the success or failure of adaptation. There is no shortage of wealth or technology in New Orleans or Victoria and yet there remain significant barriers to adaptation in these places. Conversely, there is not much wealth or technological capacity in the Sahel, in Sri Lanka or among the coffee farmers of Guatemala. Yet these cases show that factors such as the extent and quality of social capital, population mobility, an ability to learn and re-organise and the use of regulation as a policy instrument help to overcome barriers such that adaptations take place.

The cases also suggest a cautionary note: That actions that seek to reduce a risk can in fact increase it in other ways. For example, responses to the California wildfires, flood defences in New Orleans and the Mississippi and drought policy in Australia all seem to have increased the risk of disaster in the future. Our cautionary note also extends to climate change mitigation efforts, including biofuel production and hydropower development that have the potential to contribute to the amplification of climate risks for vulnerable populations.

Finally, the cases suggest there may well be limits to adaptation. Some outcomes – such as relocation from homelands in the Upper Nile or from Charleville – seem to be socially unacceptable, for while they may reduce risks in some ways, they create other kinds of risks and have unacceptable social impacts. In other cases, such as hurricane Katrina or the Victorian bushfires, the scale of change may be so large that it overwhelms engineering and management systems designed to prevent damage. It is notable that those systems that seek to resist or defend against disaster using large-scale engineering 'solutions' seem to be those most prone to fail. Those that recognise risks and seek to live with them, through adjusting planning procedures and taking into account social responses and values, seem more likely to be successful, have fewer barriers and be less maladaptive.

References

Adger, W. N. (2003). Social capital, collective action, and climate change. *Economic Geography*, 79, 387–404.

Adger, W. N. and Barnett, J. (2009). Four reasons for concern about adaptation to climate change. *Environment and Planning A*, 41, 2800–2805.

Adger, W. N., Barnett, J., Chapin, F. and Ellemor, H. (2011). This must be the place: under-representation of identity and meaning in climate change decision-making. *Global Environmental Politics*, 11, 1–25.

Adger, W. N., Dessai, S., Goulden, M. et al. (2009). Are there social limits to adaptation to climate change? *Climatic Change*, 93, 335–354.

Agrawala, S. and Fankhauser, S. (eds.) (2008). *Economic Aspects of Adaptation to Climate Change: Costs, Benefits and Policy Instruments*. Paris: Organisation for Economic Co-operation and Development.

Barnett, J. and Adger, N. (2003). Climate dangers and atoll countries. *Climatic Change*, 61, 321–337.

Barnett, J. and O'Neill, S. (2010). Maladaptation. *Global Environmental Change*, 20, 211–213.

Bigano, A., Bosello, F., Roson, R. and Tol, R. (2008). Economy-wide impacts of climate change: a joint analysis for sea level rise and tourism. *Mitigation and Adaptation Strategies for Global Change*, 13, 765–791.

Burch, S. (2010). Transforming barriers into enablers of action on climate change: insights from three municipal case studies in British Columbia, Canada. *Global Environmental Change*, 20, 287–297.

Burton, I. (1997). Vulnerability and adaptive response in the context of climate and climate change. *Climatic Change*, 36, 185–196.

Caney, S. (2010). Climate change, human rights and moral thresholds. In *Human Rights and Climate Change*, ed. S. Humphreys. Cambridge, UK: Cambridge University Press, pp. 69–90.

Donner, S., Skirving, W., Little, C., Oppenheimer, M. and Hoegh-Guldberg, O. (2005). Global assessment of coral bleaching and required rates of adaptation under climate change. *Global Change Biology*, 11, 2251–2265.

Dovers, S. (2005). *Environment and Sustainability Policy: Creation, Implementation, Evaluation*. Sydney: Federation Press.

Dovers, S. and Hezri, A. (2010). Institutions and policy processes: the means to the ends of adaptation. *Wiley Interdisciplinary Reviews: Climate Change*, 1, 212–231.

Ellemor, H. (2005). Reconsidering emergency management and indigenous communities in Australia. *Environmental Hazards*, 6, 1–7.

Farber, D. (2007). Adapting to climate change: who should pay? *Journal of Land Use and Environmental Law*, 23, 1–38.

Few, R., Brown, K. and Tompkins, E. (2007). Public participation and climate change adaptation: avoiding the illusion of inclusion. *Climate Policy*, 7, 46–59.

Gardiner, S. (2011). *A Perfect Moral Storm: The Ethical Tragedy of Climate Change*. Oxford, UK: Oxford University Press.

Grothmann, T. and Patt, A. (2005). Adaptive capacity and human cognition: the process of individual adaptation to climate change. *Global Environmental Change*, 15, 199–213.

Heyd, T. (2008). Cultural responses to natural changes such as climate change. *Espace Populations Sociétés*, 1, 83–88.

IPCC – Intergovernmental Panel on Climate Change (2007). *Climate Change 2007: Synthesis Report. Contribution of Working Groups I, II, and III to the Fourth Assessment Report of the Intergovernmental Panel on Climate Change*, eds. Core Writing Team, R. K. Pachauri and A. Reisinger. Cambridge, UK: Cambridge University Press.

Kelman, I. (2008). Island evacuation. *Forced Migration Review*, 31, 20–21.

Kuruppu, N. (2009). Adapting water resources to climate change in Kiribati: the importance of cultural values and meanings. *Environmental Science and Policy*, 12, 799–809.

Kuruppu, N. and Liverman, D. (2011). Mental preparation for climate change adaptation: the role of cognition and culture in enhancing adaptive capacity of water management in Kiribati. *Global Environmental Change*, 21, 657–669.

Leiserowitz, A., Maibach, E. and Roser-Renouf, C. (2010). *Global Warming's Six Americas. Yale Project on Climate Change*. New Haven, CT: Yale University and George Mason University.

Lorenzoni, I., Nicholson-Cole, S. and Whitmarsh, L. (2007). Barriers perceived to engaging with climate change among the UK public and their policy implications. *Global Environmental Change*, 17, 445–459.

McDonald, J. (2007). A risky climate for decision-making: the liability of development authorities for climate change impacts. *Environmental Planning and Law Journal*, 24, 405–417.

McDonald, J. (2011). The role of law in adapting to climate change. *Wiley Interdisciplinary Reviews: Climate Change*, 2, 283–295.

Mendelsohn, R. (2000). Efficient adaptation to climate change. *Climatic Change*, 45, 583–600.

Mortreux, C. and Barnett, J. (2009). Climate change, migration and adaptation in Funafuti, Tuvalu. *Global Environmental Change*, 19, 105–112.

Moser, S. (2005). Impacts assessments and policy responses to sea-level rise in three US states: an exploration of human dimension uncertainties. *Global Environmental Change*, 15, 353–369.

Moser, S. and Eckstrom, J. (2010). A framework to diagnose barriers to climate change adaptation. *Proceedings of the National Academy of Sciences of the United States of America*, 107, 22026–22031.

Næss L., Bang, G., Eriksen, S. and Vevatne, J. (2005). Institutional adaptation to climate change: flood responses at the municipal level in Norway. *Global Environmental Change*, 15, 125–138.

Nielsen, J. and Reenberg, A. (2010). Cultural barriers to climate change adaptation: a case study from northern Burkina Faso. *Global Environmental Change*, 20, 142–152.

Pickering, C. and Buckley, R. (2010). Climate response by the ski industry: the shortcomings of snowmaking for Australian resorts. *Ambio*, 39, 430–438.

Reeder, T., Wicks, J., Lovell, L. and Tarrant, O. (2009). Protecting London from tidal flooding: limits to engineering adaptation. In *Adapting to Climate Change: Thresholds, Values, Governance*, eds. N. Adger, I. Lorenzoni and K. O'Brien. Cambridge, UK: Cambridge University Press, pp. 54–63.

Schipper, L. and M. Pelling (2006). Disaster risk, climate change and international development: scope for, and challenges to, integration. *Disasters* 30(1), 19–38.

Singer, P. (2011). Changing values for a just and sustainable world. In *The Governance of Climate Change: Science Economics, Politics and Ethics*, eds. D. Held, A. Hervey and M. Theros. Cambridge, UK: Polity, pp. 144–161.

Smit, B., Burton, I., Klein, R. and Wandel, J. (2000). An anatomy of adaptation to climate change and variability. *Climatic Change*, 45, 223–251.

Storbjörk, S. (2010). 'It takes more to get a ship to change course': barriers for organizational learning and local climate adaptation in Sweden. *Journal of Environmental Policy and Planning*, 12, 235–254.

Swim, J., Clayton, S., Doherty, T. et al. (2009). *Psychology and Global Climate Change: Addressing a Multifaceted Phenomenon and Set of Challenges*. Washington, DC: American Psychological Association.

24

Lessons Learned for Adaptation to Climate Change

SARAH BOULTER, JEAN PALUTIKOF, AND DAVID JOHN KAROLY

All disasters have certain features that can be compared: The event itself, the immediate response of the community and government and the long-term response that is aimed at reducing the impact of any future similar events. The previous chapters have provided accounts of eighteen very different extreme events (summarised in Table 24.1). All, with the exception of the tsunami in south-east Asia (Hettiarachchi and Dias, Chapter 16), are directly tied to climate and weather extremes. Each presents its own unique challenges – dependant on the nature of the extreme event, how it unfolded (timing and location) and the existing vulnerability and adaptive capacity of the affected communities. Yet across the case studies, there emerge a number of common and recurring themes: First, there is a certain amount of luck or chance that determines the severity of an event (e.g. the timing or location of landfall of a cyclone determines if it is a human catastrophe or not); second, community awareness of risk plays a role (e.g. understanding that you live in a flood plain and the implications for your safety); third, communication channels play a vital role in successful management of disasters, but the infrastructure and technologies may be vulnerable; and, finally, the time factor – time can be a blessing (e.g. time to develop post-disaster adaptation) or a curse (e.g. time can erode corporate and community knowledge of risks).

In this chapter, we consider the commonalities and points of difference among the case studies in order to think about what lessons may be identified from disaster management and response to enhance our understanding of climate change adaptation.

24.1 The Events (and Before)

The scale of impact of an extreme event relates not only to its type and severity but also to its geographical location and spread, timing and location. In terms of the human impact, the existing vulnerability and preparedness of the affected community are also factors.

Table 24.1. *Summary of the extreme event case studies presented in this book*

Event	Deaths & injuries	Economic cost	Key impact	Historical context	Worsening factors	Saving grace
Hurricane Katrina (New Orleans, 2005)	1,464	Approx. US$40 billion	Flooding; breakdown of social structure	Hurricane Betsy 1965	Incomplete levee system around New Orleans	'Doomsday scenario' avoided: overtopping or breach of Pontchartrain levees could have worsened flooding; weakening of hurricane before landfall
Flooding of the Mississippi River (central United States; 1812–2005)	246 (in 1927); 385 (1937); 32 (1993)	Hundreds of millions	Flooding	Regular floods	Increasing urbanisation; levees constricting water flow	None
Wildfire in North America (southern California, 2003 and 2007)	24 (2003); 7 (2007)	US$3 billion	Property destruction, loss of forests	Nine mega fires since 1889 500 homes per year lost to fire	Exponential population growth and vulnerability of settlement lessons	Effective evacuation
Heatwaves in Kansas (US, 1980)	236 (excess deaths)	Unknown	Health, morbidity and mortality	Yes – not specified	Vulnerable community members	Development and refinement of Extreme Heat Programs
Drought in the Murray-Darling Basin (Australia)	Unknown	Unknown	Lost agricultural productivity	1895–1902 – Federation; 1937–1945 – WWII; 1997–2008 – Big Dry	Market forces; water policy	Changing policy

(cont.)

Table 24.1 (cont.)

Event	Deaths & injuries	Economic cost	Key impact	Historical context	Worsening factors	Saving grace
Victorian bushfires (Australia, 2009)	173	AU$4.4 billion (Teague et al., 2010)	Burning of houses and forested regions	1983 – Ash Wednesday; 1939 – Black Friday	Heatwave, drought, low fuel moisture, extremely low relative humidity, very strong northerly winds (forest fire danger index highest ever record – 170) Land use planning, emergency services preparedness	Effective evacuation in some areas
Cyclone Tracy (Darwin, Australia, 1974)	71 deaths; 650 injured	AU$500–600 million (1974)	94% of housing no longer habitable; 75% of population evacuated	1897, 1937	Slow translational speed of cyclone, landfall on township	Did not coincide with high tide, which would have created flooding
Queensland floods (Mackay and Charleville, 2008)	No deaths	>AU$400 million	Property and infrastructure damage	Charleville – 10 floods since 1910; Mackay floods recorded since 1884	Flash flood in Mackay	Charleville – community awareness Mackay – good emergency management support
Windstorm (Europe)	Not stated	Not stated	Property and infrastructure damage	Frequent events	Large geographic scale	Building regulations
Floods in Central Europe (Czech Republic, Poland and Germany, 1997, 2002 & 2010)	114 (1997)	€6.5 billion (1997); €15 billion (2002); US$3 billion (2010)	Flooding and inundation	Long history since 1342	Land use, changing demography and increased development; dike failure	Advance warning in Germany

North Sea tidal surge (Netherlands, 1953)	1,835 deaths, 1,916 injured	€0.7 billion (1953)	Flooding and inundation	Below sea level	Dike failure (inadequate maintenance; diffuse governance – many different small water boards were responsible for maintenance)	Further large dike on the brink of failure and saved by a ship plugging a breach
Sahel drought (North Africa, 1972–74 and mid 1980s)	Unknown	Unknown	Famine, loss of agricultural productivity, reduced growing season	Abrupt change in rainfall from 1960s	Government failure to act	In the 1980s, improved preparation by farmers and government
Indian Ocean tsunami (Sri Lanka, 2004)	Approx. 40,000	US$1 billion	Flooding, inundation, property damage, loss of life	1883 Tsunami relatively small, little damage	Flooding and inundation, building collapse. Population density, economic development and transport infrastructure concentrated along the coast. Topography and gaps in coral reef created by illegal mining. No capacity for warning.	None

(cont.)

Table 24.1 (*cont.*)

Event	Deaths & injuries	Economic cost	Key impact	Historical context	Worsening factors	Saving grace
Cyclone Sidr (Bangladesh, 2007)	3,406 deaths, 55,000 injuries	US$1.7 billion	Property damage, loss of life, loss of agriculture, loss of employment and income	One of 10 most devastating cyclones to hit Bangladesh between 1876 and 2007	Loss of shelter extremely high. Poverty.	Some evacuation
Hurricanes in Mexico and Guatemala (Mesoamerica)	Approx. 500	US$1.7 billion	Crop damage, loss of transport and infrastructure, loss of income	Mitch 1998, Stan 2005, Agatha 2010	Poverty, terrain	Community-based adaptation
Floods in the Nile Basin (Africa, 1997/1998)	Not stated	Not stated	Loss of crops and livestock, destruction of homes	1878, 1947, 1961, 1988, 1997, 1998, 2006, 2007 and 2010	Increased population since previous event	Community-based leadership, adaptation and diversification of income
Floods in China (1998, 2010)	5,500 (1998), 3,200 (2010)	US$30.7 billion (1998), US$51 billion (2010)	Flooding, inundation, loss of lives	1860, 1870, 1931, 1935, 1954, 1998	Development and urbanisation	Development of flood protection mechanisms

Across the case studies considered here, some by their nature have a widespread immediate geographical impact (e.g. flooding of the Mississippi River, heatwaves in the US and Europe and drought in Australia), while others are restricted in their impact area but with equally devastating results (e.g. cyclone Tracy in far north Australia and hurricane Katrina in the US). The timing of events can influence their impact and the actions of community members. For example, in hurricane Katrina, the call for evacuation occurred shortly before payday, and this has been cited as a reason many poorer members of the community were unable to evacuate (Garcia, Chapter 3). In the case of cyclone Tracy, the landfall early on Christmas morning meant many people were safe in their houses or away from the city for the holidays, potentially reducing the death and injury rate (Mason et al., Chapter 9). It also meant, however, that many officials were on leave, reducing the response capacity of local and regional government.

All case studies demonstrate that a community's response to and experience of an event are tied very strongly to its characteristics – demographics, economic wealth, social condition and political influence (see Huq and Ayers, Chapter 24, for further discussion of this point). Some argue that prior experience of an extreme event enhances resilience, whereas others contend that it can lead to over-confidence: Dealing with a bushfire in your 20s can be very different from coping 30 to 40 years later. Response and experience are also linked to the availability of early warning systems and whether action strategies have been developed and practiced.

Warnings are often critical in reducing the impact of extreme events. Effective capacity to warn and communicate risk often depends on community awareness that systems exist, knowledge of how to access them and the sophistication of detection and prediction systems. However, the warning in itself is insufficient – the community must also take heed and respond. In the case of hurricane Katrina, clear and accurate predictions of the storm's landfall and impact did not prevent a catastrophic human disaster (Garcia, Chapter 3). Some sectors of the community did not have the capacity to respond to the warning (e.g. the poor and disenfranchised), while the emergency management response was poorly coordinated and largely inadequate. In the case of cyclone Tracy, prior to the event, a number of false-alarm cyclone warnings appeared to dilute the community perception of risk and therefore response (Mason et al., Chapter 9). Following the horrific and catastrophic outcome of the Black Saturday wildfires in Victoria, a community alert program and warning system was adopted to prevent a reoccurrence. Yet when the warnings were activated under severe fire weather conditions, many residents did not respond (Whittaker et al., Chapter 8). Here, the perception was that the risk was insufficient to merit action, with the 'believability' of the warning linked to a need to see or smell fire. In fact, in the circumstances of this type of fire, being able to see or smell fire means it is too late to evacuate.

Even adequate warning systems and a responsive community may be overwhelmed by unprecedented conditions, leading to catastrophic disaster (e.g. Black Saturday bushfires, Australia, Chapter 8). Governments/authorities may fail to recognise a disaster unfolding, and this can be a problem both with slow-creeping events (e.g. drought) and very rapid-onset events (e.g. flash flooding, bushfire). This highlights the need to select the right indicators or triggers for warning, which, in the case of bushfires, is the type of weather conditions. Emergency management and community response need to be designed to work within the effective time frames of forecasting. Among case studies of developed nations, there appears to be a tacit reliance by communities on personal warnings – almost an expectation that someone will knock on my door if I need to evacuate.

In almost all the case studies documented here, there had been experience of historically analogous events (similar type and scale; Table 24.1), and in some cases actions to mitigate future impacts of similar events had been undertaken. In some instances, these prior 'adaptation' actions were very successful and this was clear in the outcome. For example, ongoing adjustment to heatwave events in Kansas City is continually improving the response (Mills and Snook, Chapter 6). In other cases, they were not. Despite the increasing channelisation of the Mississippi River, through increasing the height and extent of levees, these levees continue to be overtopped or breached in flood events (Kusky, Chapter 4).

24.2 Vulnerability and Resilience

Case studies of natural hazards and of community exposure demonstrate the role of vulnerability in understanding the impacts of extreme events and, in turn, in climate change adaptation planning. High vulnerability is clearly demonstrated in many of the eighteen case studies presented here, and for a variety of reasons ranging from geographical location through to population age structure. From these we can identify three broad categories of vulnerability related to socio-economic characteristics of a community, awareness and planning and preparation.

Socio-economic status of a community is often considered a key indicator of vulnerability. Certainly among the case studies, poor and disenfranchised members of a society or poorer nations in general experience high loss of life, housing and subsistence as natural hazard impacts. This is not, however, a simple developed versus developing nation division; even in the richest society, there are poorer members and areas, and poverty was a key factor in the impact of hurricane Katrina (Garcia, Chapter 3). Other socio-economic factors include political change. So, for example, in Poland a new political regime preceded the 1997 flood, and the transition of legislative structures meant that the systems in place for issuing warnings were inappropriate and cumbersome (Kundzewicz, Chapter 13).

Vulnerability will be extremely high where awareness of the risk of a particular event is low or absent, particularly among community and emergency management authorities. In the case of the 2004 Indian Ocean tsunami, while there may have been some scientific understanding of the risk, the limited history and absence of experience of such events meant that, despite a warning period, the response was limited (Hettiarachchi and Dias, Chapter 16). In the case of cyclone Tracy, building standards were not regulated for the domestic market and houses were built to meet day-to-day conditions rather than extreme events (Mason et al., Chapter 9).

In all of the case studies described in this book, weather or environmental conditions are central in shaping the experience of the hazard. But in most cases, human factors or activities act to increase or decrease risk (see Table 24.1 for 'worsening factors' and 'saving grace'). This is vulnerability, for example, the increasing risk of flooding in the Mississippi, US (Kusky, Chapter 4), Australia (King et al., Chapter 10), Europe (Kundzewicz, Chapter 13) and China (Kundzewicz et al., Chapter 20) and relates, at least in part, to increasing development. The geographical risk of floods always existed, but development, groundwater extraction and channel straightening changes to run-off pathways all change the flood risk. If the response is to build hard engineering structures such as levees, this may increase the risk, at least downstream of the flood protection. Softer solutions, for example, maintaining wetlands into which flood water can be pumped, may be more effective.

24.3 Adaptation Actions Following the Event

All the case studies described here demonstrate some level of adaptation following the event – even if these responses might be classified as 'maladaptations'. They are of two types: Immediate actions (e.g. heat warning systems, heat-stress prevention plans) and long-term change (e.g. changes and building regulations, building or retrofitting resilient infrastructure, changes to land use zoning).

Immediate Responses

In both the Black Saturday bushfire (Whittaker et al., Chapter 8) and Kansas heatwave (Mills and Snook, Chapter 6), the need for proactive emergency responses through assisted evacuation (bushfires) and intervention or physical assistance for vulnerable community members (heatwave) was identified. These adaptations recognise that community education is sometimes insufficient to evoke an appropriate community response in the face of the emergency and those people with the highest vulnerability may be unable to undertake the necessary action themselves and should be identified for physical assistance once an alert is issued.

In Sri Lanka, following the devastation of the tsunami, an all-hazards approach has been adopted as part of extreme event planning (Hettiarachchi and Dias, Chapter 16). This acknowledges that it may be difficult or even impossible to anticipate the scale of the risk of an extreme event, so that implementing a more flexible approach increases the preparedness for and resilience to the unknowns.

Long-Term Responses

Long-term responses to natural disasters may worsen existing or create new vulnerabilities, maintain the status quo or reduce the risk from future events. Following cyclones Sidr and Tracy, active policies to 'build back better' were adopted by the governing bodies. In the case of cyclone Tracy, this required an allowance of time to firstly assess the cause of building failure and then develop new building standards designed to withstand future cyclones (Mason et al., Chapter 9). The capacity for long-term recovery, however, may be stalled by pre-existing challenges. This is the case in Bangladesh, where transport and communication failures are hampering the recovery process (Paul and Rahman, Chapter 17). The experience of cyclone Tracy also highlighted that response actions can have their own risks, with the long-term evacuation necessitated by the slow start to rebuilding creating social and psychological problems and resettlement issues.

In the aftermath of a disaster, it is not uncommon for a government to undertake a process of formal or informal enquiry to determine the successes and failures of preparation and response at both government and community levels, the adequacy of forecasts and early warning systems and future actions that might reduce the damage and loss of life in future events. Examples of formal instruments include the Royal Commission following the Black Saturday fires in Australia (Whittaker et al., Chapter 8) and the Deltaworks report following the North Sea storm surge in the Netherlands (Vellinga and Aerts, Chapter 14). These are often important political instruments and facilitate the implementation of large-scale and expensive adaptations that may otherwise be politically unpalatable.

Determining the success of long-term recovery responses requires a method of evaluation. Time and repeat events are the most reliable tests of planned adaptation. In Kansas, efforts to improve the responses to severe heat events have meant an overhaul of the management by the local authorities (Mills and Snook, Chapter 6). In this case, incremental learning and testing of policies and adaptation responses in subsequent events (with a moderate frequency of reoccurrence) have been possible. A planning exercise was able to identify existing systems that could be modified or incorporated into heatwave response planning to quickly improve the response without the need for fundamental change or great expectations.

Although many of the long-term responses to disasters are engineering and planning solutions, some focus on changes to community and social structures. In the coffee-growing districts of Guatemala and flood plains of the Nile, empowering local communities through self-organisation has built resilience. While traditionally there has been separation of disaster preparedness from social-ecological development, Eakin and colleagues (Chapter 18, adaptation to tropical storms by coffee producers in Guatemala and Mexico) and Goulden and Conway (Chapter 19, floods in the Upper White Nile) demonstrate that significant gains can be made in disaster preparation by building social capital. In the case of Sahel, communities have deliberately adapted to climate variability and therefore reduced their vulnerability (Chapter 15, Batterbury and Mortimore). In this situation, adaptation can be facilitated by external actors but essentially takes place from within rather than being driven from the outside as was the case with cyclone Tracy (Mason et al., Chapter 9). An increase in GDP often follows a disaster. The investment in recovery has a positive impact on economic growth.

It often appears that political and public will is driven by the horror of an event (e.g. cyclone Tracy, Victorian bushfires, North Sea storm surge), but this must be accompanied by the capacity to act if effective outcomes are to be achieved. In the case of cyclone Sidr and the Indian Ocean tsunami, international shock and horror were reflected in the willingness to donate to the recovery, but limitations in capacity have hampered recovery efforts.

One of the greatest challenges to implementing responses to disasters is maintaining the lessons or policies developed following an extreme event in the face of day-to-day operations (e.g. see discussion in Eakin et al., Chapter 18, and the example of relaxed setback zoning in the tsunami-affected areas of Sri Lanka, Chapter 16). This problem is particularly acute where events are infrequent (e.g. tsunami) and the corporate and community knowledge of the risk potential can be lost, particularly in the face of other economic interests (e.g. urban development and population pressure) or simply as a result of migration (e.g. new residents in Queensland flood-affected areas were unaware they lived in a floodplain; King et al., Chapter 10).

24.4 Reflections on Historical Adaptation to Climate (as Different from Future Planning for Climate Change)

Recognising Change – Disaster Versus Variability

While all the case studies discussed in this book had disastrous outcomes, some were more clearly irregular extreme events while others were more regular and strongly tied to climate variability. If an extreme is identified as a risk management issue in a highly variable climate rather than an exceptional event unlikely to be repeated in the near term, the emphasis will be on building resilience and adaptive capacity rather than on disaster preparedness. Part of this may be community recognition that the presumed status quo (e.g. a stable climate) needs to be changed.

Community acceptance of change may require recognition of past mistakes (e.g. never should have built on a floodplain), which need to be addressed. Post-disaster recovery activities may provide the opportunity to correct these mistakes. For example, in the tsunami in Sri Lanka, new development boundaries were set in the immediate aftermath (although it should be noted that they were subsequently partly relaxed; Hettiarachchi and Dias, Chapter 16), and cyclone Tracy was the impetus for the development of building regulations.

In Australia's Murray-Darling Basin region, dry or drought conditions have been accepted as part of the normal climate variability in the region (Chapter 7). A shift in policy, from one of disaster response to one of building resilience and adaptive capacity, has been shaped by this growing awareness of the reality of climate variability.

Where Does Risk Lie? – Avoiding 'Maladaptation'

Actions taken in the immediate aftermath of a disaster are often directed at returning to the status quo as quickly as possible. The case studies highlight a few examples in which a clear policy to tackle failures that worsen a disaster is undertaken (e.g. building moratorium in Darwin following cyclone Tracy; 'build back better' approach following cyclone Sidr). If the recovery approach does not pause to reflect on future risks and other potential risks (e.g. the risk of a heatwave in a cyclone-prone district), then adaptive actions can create new or different risks in future events (maladaptation) or can fail to account for other threats or future changes (e.g. change in frequency or intensity). A clear example of the latter case was demonstrated in the Queensland flood chapter (Chapter 10). In both Charleville and Mackay, levee systems were constructed following earlier floods, but in the 2008 events, the floodwaters came from alternative flood sources to the river protected by the levees. In the case of Charleville, flooding came from a second tributary and, in the case of Mackay, from overland flow (i.e. flash flooding).

Arguably, the initial treatment of drought in Australia's Murray-Darling Basin as a disaster and the payment of disaster relief during drought could be seen as creating perverse outcomes (i.e. supporting continued production on very marginal farming lands) and have been a 'maladaptation' (Chapter 7). It is important that disaster responses (as with other adaptation actions) take a holistic approach and consider other interacting stressors and risks.

Acceptable Risk and the Associated Costs

In planning for disasters, most governments, communities and individuals make a decision – albeit a subconscious one in many cases – about what is an acceptable risk. However, what appeared to be an acceptable risk before an event may not seem so acceptable in the face of disaster and the political imperative to respond. The decision ultimately comes down to what (or how much) to invest for what level of risk (usually expressed as the severity of an event with a particular return period such as the 1-in-100-year event). In a number of case studies, it was apparent that the design of houses or levees was for relatively common events of low magnitude rather than extreme events (e.g. European floods, cyclone Tracy, hurricane Katrina). Faced with the aftermath of a disaster that may have a return period in the order of hundreds of years, the decision must be made: Is the cost of moving the risk threshold worth it?

This is particularly the case for centrally funded, high-cost engineered solutions such as coastal protection works (cf. privately funded but centrally regulated actions such as building code). The decision might be that emergency management actions will be relied on for very rare events rather than costly or unpopular adaptive actions.

Cost-benefit analyses of adaptation, compared to bearing the risk, are very rarely undertaken. Following the storm surge of 1953, the Dutch Government invested a staggering €9 billion in building a protection system that was designed to protect all areas against 1-in-4,000-year events, with protection for the most populated areas designed to withstand a 1-in-10,000-year event (Vellinga and Aerts, Chapter 14). These levels were based on a cost-benefit analysis and identified a much higher risk threshold than is generally the case (e.g. 1-in-100-year events for new building standards in Australia following cyclone Tracy). What is interesting to note in the action of the Dutch Government is that additional economic benefits were also gained from the plan.

The decision to undertake adaptation actions or accept an existing risk also brings with it a burden of either cost or risk and then the need to decide who will bear that cost or risk. A failure to invest in adaptation actions will devolve the potential for some risks and costs to the community, including potential loss of life, loss of property and decreased quality of life either through evacuation or temporary shelter arrangements. In the Murray-Darling Basin, a clear decision to assign risk and cost has been established, with a policy of resource users bearing the risk of climate variability and the government the cost of policy change (Botterill and Dovers, Chapter 7).

For almost all the case studies, previous events of similar scale and type had been experienced, yet in many cases, disasters unfolded with flaws in both community and emergency management response. So if history repeats, we might ask, are we failing to learn the lessons or are we tacitly accepting the risk of infrequent extreme events? Identifying how to learn adaptation lessons for climate change and extreme events remains an ongoing challenge. It will be important to maintain a perspective of risk rather than winding back policies in times of 'calm weather' or stable conditions.

24.5 The Future Impact of Climate Change on Disaster Management

Climate change is likely to require a shift in perception of natural disasters if adaptation is to be effective. Recognising the unfolding of a disaster was a problem in the lead-up to the European heatwave (Pascal et al., Chapter 12) and in some way the bushfires in Victoria, Australia (Whittaker et al., Chapter 8). In both these cases, identifying the risk of a disaster involved recognising and responding pre-emptively to climate cues. For other events, their regularity makes them no less disastrous in their impacts (e.g. drought in Australia, floods), but should they continue to be seen as disaster or part of the pattern of climate variability that may be shifting in response to climate change, thus necessitating a change in preparedness and response strategies? This key question has important political and policy implications. If an event is seen as part of a relatively stable pattern of natural variability, then effort can be focused on how to manage actions for sustainably dealing with future rare events. If, however, it is seen as part of a shift to a new condition, this would imply a new response with, for example, investment in development in areas not prone to flooding or moving existing flood-prone development out of high-risk areas. Expenditure will be channelled into adaptation rather than post-disaster 'clean-up' funding. This demonstrates how a shift in thinking might be undertaken that shifts the 'acceptable risk' boundary.

Under climate change, adaptive planning for extremes may require consideration of worsening scenarios. The Victorian Black Saturday bushfires in 2009 were the worst fires ever experienced in Australia and were strongly linked to very extreme climate conditions (Whittaker et al., Chapter 8). Ultimately, existing policies, strategies and responses were inadequate for the escalation of conditions. During an extensive process of enquiry through a Royal Commission, no real consideration was given to climate change and the potential for even worse conditions.

A challenge for climate change adaptation may be the absence of public 'horror' at the incremental rate of increase in severity and frequency of extremes. Lessons from gradual-onset disasters such as drought and heatwaves may offer more profound insights into climate change adaptation.

24.6 Lessons Learned

After a natural disaster, particularly one in which lives are lost, a process of investigation, enquiry and reflection generally occurs. It happens at the level of the individual, the larger community and the government. The purpose of this reflection is usually to ensure the scale of the disaster is not repeated and often results in changes – including to individual actions (e.g. encouraging people to plan their response in advance of an event, including early evacuation), as well as community actions, government policy and infrastructure. It is from this reflection that emergency management builds and refines its policies and actions, builds resilience and reduces risk. The synergies with climate change adaptation are clear, with the key tenets of adaptation being to develop adaptive capacity, resilience and adaptation frameworks/policies. The following are lessons we have derived from the historical case studies in this book, which can be translated into lessons for climate change adaptation.

Community Awareness

In all case studies, there was a considerable potential role for community awareness in addition to any government or authority role for action, enforcement or assistance. Realising this potential has a positive effect on the outcome of an event, saving lives and property. The concept of preparedness is not new in disaster management, but when shifted to the context of climate change adaptation, it has a new character.

Communities that live in an area of, for example, fire risk need to be aware of the risk as well as be prepared through a fire plan and evacuation strategy. Awareness leads to timely implementation of the plan and evacuation. Conversely, as Whittaker and colleagues demonstrate in the 2009 Black Saturday bushfires (Chapter 8), lack of awareness may result in costly inaction or delay.

In adapting to climate change, awareness extends to the realisation that something has changed and that current levels of preparedness and responses may no longer be appropriate. This might mean only minor changes but may extend to fundamental reconsideration of, for example, whether wholesale relocation of a community would be appropriate. In the case of the extraordinary flash flood in the small Australian town of Grantham, there was a post-disaster resolution by the local council to offer at-risk residents a land swap, moving them to higher ground to avoid any risk of a repeat catastrophe (Karoly and Boulter, Chapter 25).

The Whole Picture: All Hazards

In responding to disasters, it is important that actions be pragmatic, realistic and forward thinking. This is equally important for adaptation actions. The key risk of not undertaking a forward-thinking approach is that unintended consequences can result from short-term solutions so that these may not prove to be suitable in the long term (e.g. disaster payments for drought in Australia, Chapter 7). In addition, solutions that look only to the most recent disaster or to one risk factor can result in unintended consequences for future events or other areas. For example, the development of cyclone-resistant housing standards could create housing that is unsuitable for extreme heat conditions. In adaptation terms, this could be considered a maladaptation.

Enabling Adaptation

One of the greatest challenges for public policy in considering climate change is how to make adaptation happen. The case studies discussed in this volume demonstrate quite clearly that there are several pathways to enabling adaptation. In particular, adaptation can be fostered through both incentives and regulation. In the case of cyclone Tracy, the use of regulation was the most expedient approach, changing building regulation to ensure housing met cyclone standards (Mason et al., Chapter 9). For other adaptation actions, incentives may be more appropriate (e.g. buy-backs to encourage people to leave areas of extreme bushfire risk).

There is also a very clear role for community knowledge and communication. The approach undertaken by the Kansas City Council to manage heatwave exposure demonstrates a very successful communication strategy, supported by proactive actions targeted at protecting the most vulnerable citizens (Mills and Snook, Chapter 6).

Practicality

Vulnerability, while able to be minimised and countered through adaptive capacity, it can be unavoidable. For example, vulnerability can be tied to geography – remote location, exposure to extreme events and so forth. This may sometimes mean a change in community perception or attitude – for example, an acceptance that it is no longer appropriate to build in a particular locality that is regularly flooded. In the case of developing countries, vulnerability is often tied to socio-economic factors such as underdeveloped infrastructure, poverty or limited education. Adaptation actions may therefore need to address some of these issues in order to improve adaptive capacity.

Final Reflections

In observing the actions of some communities or authorities following an extreme event, the question can be asked whether those actions are truly developing resilience or are simply a reflection of stoicism. The concept of resilience embraces an element of change. Returning a community to its pre-disaster state is not always optimal but is sometimes perceived within a community and the popular media as demonstrating worthy resilience. This view can be a barrier to adaptation, as it reinforces the position of those within society who are reluctant to change or implement new strategies. Those who do seek change, for example by selling the farm or relocating their home, can be labelled as 'giving up'. All this militates against positive action and change.

Few of the case studies presented here are unique events, and many share outcomes and critical lessons. So the question can be asked: If we know many of the consequences of these disasters, why don't things change? Although in some case studies there are changes, in fact many of these are incremental, rarely transformational. The response to the 1953 storm surge might be considered an exception, with a massive investment in major infrastructure to protect communities from a repeat disaster (Vellinga and Aerts, Chapter 14). Clearly, there are barriers to translating knowledge into action, particularly transformed action, and these barriers are important to understand and inform planning for adaptation to climate change.

With every disaster comes cost – lost lives, repair or replacement of infrastructure, clean-up costs. With increasing risks of climate change–related impacts, the cost of dealing with disasters is likely to rise further unless a more resilient society and environment can be constructed. Costs will potentially become prohibitive, and ultimately the cost, if government action or preparation is not taken or fails, falls onto the community through loss of life, loss of shelter and loss of income. Most governments are unwilling to be perceived as or to become the 'insurer of last resort'. They are moving towards resilience-building strategies as an approach to disaster management (e.g. Australia's National Disaster Resilience Framework). But this also can be seen as shifting responsibility and costs from governments to communities. In planning climate change adaptation, costs are inevitably a major consideration: What is considered an acceptable risk; are we willing to pay; is the investment in adaptation worthwhile given the potential cost of not adapting? Government, while reluctant to be the 'insurer of last resort', does arguably have a role in actions that are in the interests of the public good. Government decisions to invest in adaptation may reduce the cost of rectifying damage – be that from a disaster or through a changing climate.

Protection of life is generally through early warning systems, community awareness and action and assistance to the most vulnerable. Protection of property relates to understanding what an acceptable risk threshold is and regulating to ensure structures and development meet that risk level.

Adaptation in a Changing Climate

In this volume of case studies, we have considered the development of adaptation following natural hazards. Climate change will potentially bring with it more disasters but also more gradual but equally difficult changes that must be adapted to. From the response to natural disasters, we begin to understand the process and long-term outcomes of undertaking adaptation actions. It highlights the challenge of developing political and community will to both acknowledge present and future change and implement adaptation actions and policies to address these changes. It demonstrates the potential for 'calm weather' periods to undermine strategic policy – although under climate change, we expect this will be less of a challenge than initiating adaptation policies. It also highlights the important role of cost-benefit analysis in guiding the development of appropriate adaptation strategies.

While the need to invest in adaptation for climate change can be financially challenging and politically unpalatable, disasters demonstrate the great risks that can be associated with failing to adapt. The critical difference under climate change is that a return to normal or stable conditions following an event will not occur under climate change.

Reference

Teague, B., McLeod, R. and Pascoe, S. (2010). *2009 Victorian Bushfires Royal Commission Final Report: Summary*. Victoria: State of Victoria.

25

Afterword: Floods, Storms, Fire and Pestilence – Disaster Risk in Australia During 2010–2011

DAVID JOHN KAROLY AND SARAH BOULTER

Most of the natural disasters described in this book were single events affecting one region at one time. They were considered usually as isolated or individual events in the discussion of responses. However, climate change acting together with other factors, including population growth and coastal development, is likely to increase the risk of multiple natural disasters occurring in a short period of time and affecting the same region or country. Nearly simultaneous natural disasters will impose much greater stresses on emergency management systems than isolated events, stretching resources and systems.

So, what happens when multiple natural disasters affect the same country over a short period? In late 2010 and the first half of 2011, different parts of Australia experienced a wide range of natural disasters (or risk of disasters), with locust plagues in south-eastern Australia; wildfires and heatwaves in Perth in south-west Australia and Sydney on the east coast; flooding in many parts of northern, eastern and south-eastern Australia; severe thunderstorms causing flash flooding in south-east Queensland; and a Category 5 tropical cyclone making landfall on the north-east coast (Figure 25.1). While these were not all the biblical plagues of Egypt, the country could be forgiven for thinking 'the gods' were angry.

In some of these events, good preparation and emergency management meant almost no loss of life and actual damage much less than the potential (tropical cyclone Yasi in Queensland, locust plague in south-eastern Australia). Elsewhere, systems and procedures were inadequate to prevent loss of life because of flash flooding (south-east Queensland). Here we review the unfolding of events and cumulative impact of these disasters.

25.1 Disasters and Response – Australia in 2010/2011

25.1.1 Locust Plague

As early as autumn of 2010, the Australian Plague Locust Commission (APLC) and governments of South Australia, Victoria and New South Wales (NSW) identified

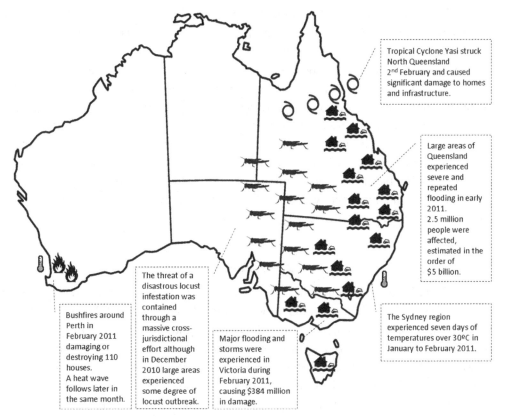

Tropical Cyclone Yasi struck North Queensland 2nd February and caused significant damage to homes and infrastructure.

Large areas of Queensland experienced severe and repeated flooding in early 2011. 2.5 million people were affected, estimated in the order of $5 billion.

The threat of a disastrous locust infestation was contained through a massive cross-jurisdictional effort although in December 2010 large areas experienced some degree of locust outbreak.

Bushfires around Perth in February 2011 damaging or destroying 110 houses. A heat wave follows later in the same month.

Major flooding and storms were experienced in Victoria during February 2011, causing $384 million in damage.

The Sydney region experienced seven days of temperatures over 30ºC in January to February 2011.

Figure 25.1 Schematic map of disaster risks during the summer of 2010–2011 showing the widespread geographical impact of the events on Australia.

the potential for a major spring infestation of the Spur Throated Locust in southern Australia. The tropical Queensland locust species rarely reaches plague proportions in southern states unless wet, humid conditions persist. In 2010, heavy summer rainfall boosted locust numbers with high autumn laying and a southward migration of adults. With high numbers of egg-laying adults, a high spring hatching was predicted as a result. A coordinated effort to prepare for the outbreak and minimise damage was initiated by state authorities and the APLC (APLC, 2010; Millist and Abdalla, 2011).

With at least six months in which to prepare for the outbreak, considerable resources were poured into stockpiling chemicals, preparing staff and equipment, surveying 'hotspots', and raising community knowledge and awareness. Regional and local preparations also included financing chemical rebates to landholders and the establishment of regional stakeholder groups. Locusts were targeted during the nymphal or band stage, before they reached the far more difficult-to-control stage in which they fly and form swarms (APLC, 2010).

A cost-benefit analysis of the operation estimated that the potential losses of the locust outbreak were in the order of AU$963 million. Investment in treatment by all parties was in the order of AU$50 million, giving a net saving of AU$913 million (Millist and Abdalla, 2011), although significant (unquantified) damage still occurred across the affected states (Ludwig, 2011).

The success of the control operation appears to be because of the early activation of response and planning, an appreciation of the potential severity of an uncontrolled outbreak, investment and cooperation across several jurisdictions, establishment of community and regional reference groups that allowed communication between local landholders and operational people on the ground as well as a strong community awareness campaign. In South Australia, for example, Biosecurity SA undertook a six-month campaign that harnessed community engagement, media liaison, fact sheets, a constantly updated website, weekly e-news and SMS text messages to subscribers for aerial spraying alerts (Biosecurity SA, 2010).

25.1.2 Floods

Beginning in Victoria in September 2010 (Comrie, 2011), widespread flooding was experienced across eastern Australia through until February 2011 (BOM, 2011a). The state of Queensland experienced some of the worst flooding, affecting three-quarters of the state and 2.5 million people. Thirty-five people lost their lives, with a further three missing (QFCI, 2011). The Australian Bureau of Meteorology recorded flood peaks at more than 100 Queensland river-height stations, and some 29,000 homes and business experienced some level of inundation. The cost of recovery has been estimated on the order of AU$5 billion (Queensland Government, 2011). Major flooding was also experienced in NSW, Victoria and Tasmania at various times (BOM, 2011a).

At the height of the flooding in Queensland (and leading into the approach of cyclone Yasi, which made landfall on 3 February), the state's premier became the primary source of up-to-date and accurate public information through regular briefings, advising her constituents, 'If you hear or read on social networking sites rumours or statements, if they're not confirmed in these meetings, if you don't hear them out of my mouth or out of the police commissioner or deputy commissioner then it's very unlikely to be true' (Bligh, 2011).

Recovery

In response to widespread, regional flooding, the Queensland government established the Flood Recovery Taskforce on 5 January 2011. The scope of this body continued to expand as the scale of the disaster increased, and its role included the post–cyclone Yasi recovery efforts. A massive volunteer effort was also involved in the clean-up of affected suburbs in the cities of Brisbane and Ipswich.

In the aftermath of the flooding, a number of formal inquiry processes were set up to examine planning and response measures with the view to improve these in future events. In Queensland, the Queensland Floods Commission of Inquiry released an interim report in August 2011 that made 175 recommendations on matters of flood preparedness, with the intent that these be available and actioned prior to the next wet season (QFCI, 2011), and a final report in March 2012 that made further comments on planning, development, mining, infrastructure, the operation of the Wivenhoe Dam and the deaths (QFCI, 2012). While a clear public warning and response effort was very successfully conducted, there were a number of aspects of the management of events that have attracted considerable public scrutiny: The limited warning of the most catastrophic flash flooding event in the Toowoomba and great Lockyer Valley region and the role of dam operations in the flood levels experienced in Ipswich and Brisbane (QFCI, 2012). The operation of the dam came under considerable scrutiny. The floods were preceded by several years of severe drought, and it became apparent that the procedures for flood mitigation conflict with the requirements for storing sufficient water in drought conditions for a large urban centre. The operations manual of the dam reflected conflicting procedures.

A similar inquiry process in Victoria examined the response to flooding in that state. In this case, the Victorian Flood Review (VFR) identified the same shortcomings in the state's emergency management arrangements that had been identified in the Victorian Bushfire Royal Commission (see Whittaker et al., Chapter 8). They noted that, despite the appropriateness of an 'all hazards, all agencies' philosophy of emergency management, it was not effectively operationalised because of barriers created by the organisational culture and lack of communication, coordination and information sharing. Critically, a lack of robust policy that could facilitate a coordinated or adequate command structure meant the response to the floods was *ad hoc*. One reason for these shortcomings was the adherence to artificial administrative boundaries, whereas a regional-scale response with clear articulation of roles would have ensured best practice in flood warning. The review also raised serious concerns over the vulnerability of privately owned and operated infrastructure in floodplains, the message being that mitigation of flood risk should be incorporated in planning and development (Comrie, 2011).

Rebuilding and Adapting

In one of the most severely affected townships, Grantham, the local council took the extraordinary step of relocating residents from the valley bottom affected by the floods to higher land. A very rapid response to the devastation saw the concept of relocation mooted in the first week after the floods. The local council prepared a community recovery and relocation plan together with a new master plan for the town, which moved at-risk residents to safer ground. In a fast-tracked development, residents were able to apply for the voluntary land swap in June 2011, with a ballot in August giving about 50 per cent of applicants their first preference. Residents began moving in at the end of 2011. This move affects about one-third of the population of Grantham, and there is no doubt that it is key to the long-term viability of the community, given the devastation wrought by the floods. The entire project is expected to cost the council AU$30 to 40 million.

Across Queensland, new mapping of floodplains is being undertaken to assist councils in planning and development. The Queensland Flood Commission of Inquiry considered the issue of zoning, although no recommendations were made in its final report.

In Brisbane, the local council introduced temporary planning provisions as of 16 May 2011 aimed at reducing flood damage risk to new housing and rebuilding of damaged housing in flood prone areas. In advice to residents wishing to flood-proof rebuilding efforts, the council suggests, 'You could consider raising your house ... and use building products that have higher water resistance ratings' (BCC, 2011), with the recommendation that habitable areas of a house should be raised 500 mm above 2011 flood levels. In areas where flood levels already reached the second storey of some properties, this may not be a practical option. This then raises the long-term issue of whether there should be some program of buy-back in very vulnerable locations.

25.1.3 Cyclone Yasi

Severe tropical cyclone Yasi crossed the coast of north Queensland on 3 February 2011 as a Category 5 storm. Warning of the system and its potential to develop into a very large and severe event was available from as early as 28 January (Withey and Bavas, 2011). The government reactivated the State Disaster Management Group (recently stood down following the Queensland flooding) to prepare for the storm. Considerable effort was made to understand the potential for damage and storm surge, and a number of precautionary evacuations were instigated, including the complete evacuation of Cairns's two hospitals.

Although the storm brought with it extensive rainfall, wind damage and surge, there was only one recorded death (because of asphyxiation from diesel generator fumes). Significant wind damage was reported between Innisfail and Townsville (a distance of approximately 200 km; Boughton et al., 2011). The Insurance Council of Australia estimated the 2011 damage bill at AU$1.4 billion (Insurance Council of Australia, 2012).

Again, a very clear warning and message was made to residents, with the state's premier identifying the window of opportunity to evacuate and advising residents on the best preparations and most sheltered place to stay in their houses in the lead-up to cyclone landfall.

Rebuilding is still ongoing two years after cyclone Yasi crossed the coastline. Significant investment was made in both the clean-up and improvements for future resilience. The Queensland Reconstruction Authority (QRA) undertook a program of 'build back better' with reconstruction to current engineering standards as a minimum, and as described by the QRA Chairman, where it 'makes sense', improvement of the rebuilt structure (van Vonderen, 2012).

For the state of Queensland, the succession of disasters across several years has seen the reconstruction authority take on a more permanent ongoing role and has been instrumental in developing strategies for disaster resilience (e.g. Queensland Reconstruction Authority, 2012).

25.1.4 Heatwaves and Fire

The Sydney region experienced seven days of temperatures higher than 30 °C from 31 January 2011, with a maximum of 41.5 °C being reached on 5 February (BOM, 2011b). This event featured the hottest night on record, of 27.6 °C. Night-time temperatures were exceptionally warm and coincided with high humidity associated with recent tropical cyclone activity (T.C. Anthony and Yasi). On 6 February, the NSW Health Department said sixty-two people had been treated in emergency departments for heat-related illness in the preceding six days (O'Rourke et al., 2011), and throughout the week, paramedics responded to 213 cases of heat-related illness across the state (Schwartzkoff, 2011). Although electricity demand reached peak levels, with a record weekend demand on 5 February, only minimal power outages were experienced (Schwartzkoff, 2011). On 31 January, NSW Health issued warnings to communities about how to behave and take care of themselves in heatwave conditions.

In the same period, extreme fire weather was being experienced in Western Australia. A total fire ban was invoked in the Perth region on 3 February 2011. The first fires began on 5 February after a prescribed burn escaped containment lines, spreading very rapidly and eventually destroying seventy-one homes and damaging a further thirty-nine. Fortunately, no lives were lost (Keelty, 2011).

Early warnings of the impending severe fire season were issued by the Bureau of Meteorology (BOM) in October of 2010, and the WA Minister for Emergency Services was able to advise the WA parliament of the severity of the risk (Emerson, 2010).

25.2 The National Scale

This series of natural disasters across Australia in 2010 and early 2011 was both directly and indirectly very costly to the nation. Insurance claims across the five events equal some AU\$4.4 billion, while estimated government expenditure on the recovery is set to exceed AU\$12 billion (Table 25.1). To meet the huge demand for rebuilding costs, the Australian government introduced legislation for a once-off tax levy (Gillard, 2011).

The costs to government budgets were also felt through a number of market mechanisms. In the first quarter of 2011, the Australian GDP shrank by 1.2 per cent

Table 25.1 *Cost of the disasters*

Event	Date	Location	Insurance claims (AUD)[1]	Government expenditure (AUD)
All natural disasters	2010/2011 financial year		N/A	$6.6 billion (Federal)[2]
Locust outbreak	2010/2011	Victoria, South Australia, NSW	N/A	$50 million
Queensland flooding	21/12/2010 to 14/01/2011	Queensland, rural, Toowoomba, Lockyer Valley	$2.39 billion	$5 billion (Queensland)[3]
Victorian flooding	12/12/2011 to 18/1/2011	Victoria	$126 million	$676 million (Victoria)[4]
Cyclone Yasi	2/2/2011 to 7/2/2011	Far north Queensland	$1.41 billion	(included in total for Qld flooding)
Severe storms Victoria	4/2/2011 to 6/2/2011	Victoria	$488 million	
Perth Hills bushfires	5/2/11 to 7/2/11	Perth and surrounds	$35 million	
Totals			**$4.45 billion**	**$12.33 billion**

(seasonally adjusted), and this was largely attributed to the impact of natural disasters, in particular the widespread flooding (Smith, 2011). Budget forecasting predicted reductions in tax receipts as a result of businesses being affected by disasters. For example, coal exports fell 25 per cent after the floods, wiping out AU$7 billion worth of coal production in Queensland alone (Smith, 2011). The value of the Australian dollar was particularly high during the period following the floods, and it has been suggested that one reason for this is the increased demand by foreign reinsurers for Australian dollars to meet pay onto the insurance companies (Glynn, 2011).

National Emergency Management Planning

In 2009, the Council of Australian Governments released a national disaster resilience statement. This document sets out a direction for emergency management to build national resilience. The policy signalled a shift from focussing on documenting roles, responsibilities and procedures by building upon existing emergency planning

---

[1] Insurance Council of Australia as at 21 January 2013.
[2] Federal budget predictions over next six years.
[3] Queensland Government, 2011.
[4] Victorian state budget released in May 2011 (AU$115 million expected to be recovered from insurance).

arrangements to focusing on action-based resilience planning to strengthen local capacity and capability, improving community engagement and a better appreciation of community diversity, needs, strengths and vulnerabilities (NEMC, 2009). A national emergency alert system has been established in Australia, which sends recorded messages to phones based on billing address in the case of an emergency (see www. emergencyalert.gov.au).

25.3 The Role of Climate

Large-scale patterns or modes of climate variations, such as El Niño-Southern Oscillation affecting countries bordering the Pacific Ocean or the North Atlantic Oscillation, tend to organise climatic extremes so that multiple natural disasters are more common in some years in some regions. A very strong La Niña event occurred in 2010 and 2011, with colder-than-normal ocean temperatures in the equatorial Pacific Ocean and associated increases in rainfall over eastern Australia and reduced rainfall on the Pacific coast of tropical South America (Blunden and Arndt, 2012).

The typical pattern of rainfall anomalies in Australia during a La Niña event was very strong in 2010/11, with record high rainfall in a number of parts of northern, eastern and southern Australia but record low rainfall in south-west Australia (Figure 25.2; Ganter and Tobin, 2012). La Niña events are also associated with an increased risk of land-falling tropical cyclones on the north-east coast (Evans and Allan, 1992) and above-normal temperatures in south-west Australia (Jones and Trewin, 2000). Hence, La Niña provided the large-scale climatic environment that increased the chances of these natural disasters affecting Australia in the same period. In addition, there were record high sea surface temperatures in the eastern Indian Ocean to the north-west of Australia, which likely enhanced the typical La Niña pattern. The previous very strong La Niña in 1974 was also associated with flooding in Brisbane and much of eastern and southern Australia, as well as the severe tropical cyclone Tracy that caused extensive damage to the city of Darwin in northern Australia.

While the possible contribution of anthropogenic climate change to extreme weather and climate events in 2011 in other regions has been considered (Peterson et al., 2012), no such studies have been completed for these extreme events in Australia yet.

25.4 Conclusion

A number of the disasters reported here have been followed up by a formal inquiry process that assesses the response of community and emergency management, considers existing vulnerabilities and makes recommendations for actions to reduce the impact of future events (adaptations). This model has been employed in Australia

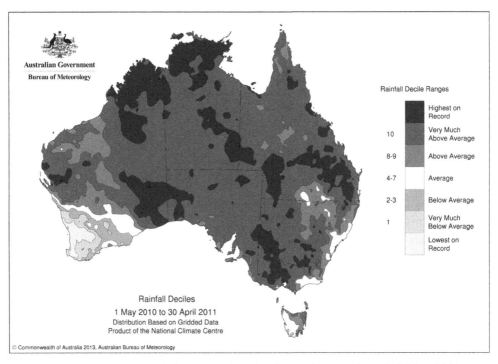

Figure 25.2 Australian rainfall anomalies for the twelve-month period May 2010–April 2011, showing large parts of central and eastern Australia experienced record high rainfall but the south-west of Western Australia experienced record low rainfall (Australian Bureau of Meteorology).

following some its worst disasters (e.g. the Black Saturday fires described in Chapter 8). In Queensland, the reconstruction and recovery effort is aimed at rebuilding a 'stronger, more resilient Queensland' (Queensland Reconstruction Authority, 2011). It is clear that the detailed assessment of the successes and limitations of past emergency management responses can help to build improved responses to future events.

However, while one of the goals of such inquiries is to aid in the rapid rebuilding of safe communities and to minimise the risk of future natural disasters, sometimes their limited scope can lead to perverse outcomes. The IPCC Special Report *Managing the Risks of Extreme Events and Disasters to Advance Climate Change Adaptation* (IPCC, 2012) makes it clear that effective management of the risks of weather and climate-related disasters requires the consideration of anthropogenic climate change as a one of the multiple stressors increasing the risks of extreme events. This has not always been the case in the recent inquiries in Australia.

Multiple natural disasters affecting one country over a relatively short period, such as Australia in 2010/11, provide an indication of the limitations of natural disaster management systems in coping with current extreme events and the even greater limitations likely under the compounding effects of future climate change.

In particular, the structural organisation of the responding organisations appears to fall short of what is necessary to deal with large-scale and multiple rapid-succession disasters.

Unlike the biblical plagues of Egypt, there is no book with which to inspire 'belief' in climate change, and one of the greatest challenges for climate change adaptation is convincing the wider community and the political body of the need for action now in response to a growing threat. In Australia, there appears to be an alarming trend toward increasing scepticism despite the daily growth in scientific literature that shows the future may be grimmer than we first thought. Human nature means we understand and respond to immediate or previously experienced threats, but if we can't make an emotional connection to a problem, we defer action. The experience of disasters may be the best 'surrogate' or 'analogue' we have to the problems of and pathways to adapting to climate change.

In Australia, a strong community program of risk awareness using imagery of king tides (as an analogue for sea-level rise) provides community-based guidance to build resilience (see www.greencrossaustralia.org). The 'Harden Up' slogan adopted for this Web-based agent-of-change program suggests to the community that they must empower themselves, but the website also uses disasters and extreme events as a pathway to develop community resilience and adaptation action. The lessons provided by the natural disasters described in this book can assist with adaptation to climate change, but only if we listen.

References

APLC – Australian Plague Locust Commission (2010). General situation in April and outlook to spring 2010. *Australian Plague Locust Commission Bulletin*, May 2010. Accessed 17 December from: <http://www.daff.gov.au/_data/assets/pdf_file/0011/1623656/bulletin-may2010.pdf>.

BCC – Brisbane City Council (2011). *Floods Fact Sheet Advice for Residents. Approvals for Repairing, Renovating or Rebuilding Flood Affected Houses*. Brisbane: Brisbane City Council. Accessed 17 December 2012 from: <http://www.brisbane.qld.gov.au/2010%20Library/2009%20PDF%20and%20Docs/5.Community%20Support/5.4%20Emergency%20management/Factsheet_Planning_and_Building_Residents.pdf>.

Biosecurity SA (2010). *Locust Management in Grain Crops and Pastures*. Adelaide: Primary Industries and Regions South Australia. Accessed 10 June 2013 from: <http://www.pir.sa.gov.au/_data/assets/pdf_file/0019/131059/locstman_Grain_Crops_and_Pastures_ds.pdf>.

Bligh, A. (Premier of Queensland) (2011). *Transcript – Press Conference – 11:30 am Thursday, 13 January 2011*, media statement. Accessed 17 December 2012 from: <http://statements.qld.gov.au/Statement/Id/73283>.

Blunden, J. and Arndt, D. S. (2012). State of the climate in 2011. *Bulletin of the American Meteorological Society*, 93(7), S1-S264.

BOM – Bureau of Meteorology (2011a). *Frequent Heavy Rain Events in Late 2010/Early 2011 Lead to Widespread Flooding Across Eastern Australia. Special Climate Statement 24*. Melbourne: National Climate Centre, Bureau of Meteorology. Accessed 10 June 2013 from: <http://www.bom.gov.au/climate/current/statements/scs24c.pdf>.

BOM – Bureau of Meteorology (2011b). *Sydney in Summer 2010–11: A Dry Summer with Very Warm Nights*. Sydney: Bureau of Meteorology. Accessed 18 December 2012 from: <http://www.bom.gov.au/climate/current/season/nsw/archive/201102.sydney.shtml#recordsTminDailyHigh>.

Boughton, G. N., Henderson, J. D., Ginger, J. D. et al. (2011). *Tropical Cyclone Yasi Structural Damage to Buildings. CTS Technical Report No 57*. Townsville: Cyclone Testing Station, James Cook University.

Comrie, N. (2011). *Review of the 2010–11 Flood Warnings & Response. Interim Report by Neil Comrie AO, APM, 30 June 2011*. Melbourne: State of Victoria.

Emerson, D. (2010). Grim outlook for bushfires this summer. *The West Australian*, 12 November 2010. Accessed 18 December 2012 from: <http://au.news.yahoo.com/thewest/a/-/wa/8305010/grim-outlook-for-bushfires-this-summer/>.

Evans, J. L. and Allan, R. J. (1992). El Niño/Southern Oscillation modification to the structure of the monsoon and tropical cyclone activity in the Australasian region. *International Journal of Climatology*, 12, 611–623.

Ganter, C., and Tobin, S. (2012). Australia [State of the climate in 2011]. *Bulletin of the American Meteorological Society*, 93, S218-S221.

Gillard, J. (Prime Minister of Australia) (2011). *Rebuilding after the Floods*, media release, Commonwealth of Australia, Canberra, 27 January. Accessed 12 December 2012 from: <http://www.pm.gov.au/press-office/rebuilding-after-floods>.

Glynn, J. (2011). Reinsurance costs in Yasi, Queensland floods trigger Australian dollar surge. *The Australian*, 31 March 2011. Accessed 18 December 2012 from: <http://www.theaustralian.com.au/business/markets/reinsurance-costs-in-yasi-queensland-floods-trigger-australian-dollar-surge/story-e6frg94o-1226031290450>.

Insurance Council of Australia (2012). *Catastrophe Events and the Community*. Sydney: Insurance Council of Australia. Accessed 18 December 2012 from: <http://www.insurancecouncil.com.au/issues-submissions/issues/catastrophe-events>.

IPCC – Intergovernmental Panel on Climate Change (2012). *Managing the Risks of Extreme Events and Disasters to Advance Climate Change Adaptation. A Special Report of Working Groups I and II of the Intergovernmental Panel on Climate Change*, eds. C. B. Field, V. Barros, T. F. Stocker et al. Cambridge University Press: Cambridge, UK.

Jones, D. A. and Trewin, B. C. (2000). On the relationships between the El Niño-Southern Oscillation and Australian land surface temperature. *International Journal of Climatology*, 20, 697–719.

Keelty, M. J. (2011). *A Shared Responsibility. The Report of the Perth Hills Bushfire February 2011 Review*. Perth: Government of Western Australia.

Ludwig, J. (Minister for Agriculture, Fisheries and Forestry) (2011). *ABARES Study Finds Locust Dollars Well Spent*, media release, Commonwealth of Australia, Canberra, 25 March. Accessed 12 December 2012 from: <http://www.maff.gov.au/media_office/media_releases/media_releases/2011/march/abares_study_finds_locust_dollars_well_spent>.

Millist, N. and Abdalla, A. (2011). *Benefit-Cost Analysis of Australian Plague Locust Control Operations for 2010–2011. ABARES Report Prepared for the Australian Plague Locust Commission*. Canberra: Australian Bureau of Agricultural and Resource Economics and Sciences.

NEMC – National Emergency Management Committee (2009). *National Strategy for Disaster Resilience*. Canberra: Council of Australian Governments.

O'Rourke, J., Dingwall, D. and Barlass, T. (2011). City seeks refugee from heat. *The Sydney Morning Herald*, 6 February 2011. Accessed 18 December 2012 from: <http://www.smh.com.au/environment/weather/city-seeks-refuge-from-heat-20110205-1ahmj.html>.

Peterson, T. C., Stott, P. A. and Herring, S. (2012). Explaining extreme events of 2011 from a climate perspective. *Bulletin of the American Meteorological Society*, 93, 1041–1067.

QFCI – Queensland Floods Commission of Inquiry (2011). *Queensland Floods Commission of Inquiry Interim Report*. Brisbane: Queensland Floods Commission of Inquiry.

QFCI – Queensland Floods Commission of Inquiry (2012). *Queensland Floods Commission of Inquiry Final Report*. Brisbane: Queensland Floods Commission of Inquiry.

Queensland Government (2011). *Queensland State Budget 2010–11. Mid Year Fiscal and Economic Review*. Brisbane: Queensland Government.

Queensland Reconstruction Authority (2011). *Rebuilding a Stronger, More Resilient Queensland*. Brisbane: Queensland Government. Accessed 18 December 2012 from: <http://www.qldreconstruction.org.au/u/lib/cms/Resilience-doc.pdf>.

Queensland Reconstruction Authority (2012). *Operation Queenslander: The State Community, Economic and Environmental Recovery and Reconstruction Plan 2011–2013 Update*. Brisbane: Queensland Government. Accessed 18 December 2012 from: <http://www.qldreconstruction.org.au/u/lib/cms2/operation-queenslander-state-plan-full.pdf>.

Schwartzkoff, L. (2011). Finally, some relief after hottest night in history. *The Sydney Morning Herald*, 7 February 2011. Accessed 18 December 2012 from: <http://www.smh.com.au/environment/weather/finally-some-relief-after-hottest-night-in-history-20110206-1aif4.html>.

Smith, P. (2011). Natural disasters hit Australia's GPD. *Financial Times*, 1 June 2011. Accessed 18 December 2012 from: <http://www.ft.com/cms/s/0/24992bf0-8bfe-11e0-854c-00144feab49a.html#axzz2FS7uep8G>.

van Vonderen, J. (2012). Reconstruction authority chairman says Yasi rebuilding on track. *ABC News*, 3 February 2012. Accessed 23 January 2013 from: <http://www.abc.net.au/news/2012–02–03/reconstruction-authority-chairman-says-yasi/3811360>.

Withey, A. and Bavas, J. (2011). North Qld on cyclone alert again. *ABC News*, 28 January 2011. Accessed 18 December 2012 from: <http://www.abc.net.au/news/stories/2011/01/28/3123758.htm?site=tropic>.

Index